Grade 2

Reveal MATH®

Assessment Resource Book

Mc
Graw
Hill

mheducation.com/prek-12

Send all inquiries to:
McGraw Hill
8787 Orion Place
Columbus, OH 43240

ISBN: 978-1-26-421056-5
MHID: 1-26-421056-6

Printed in the United States of America.

9 10 11 12 LON 27 26 25 24

Grade 2
Table of Contents

Unit **1**
Math Is …

Unit **2**
Place Value to 1,000

Unit 3
Patterns within Numbers

Unit 4
Meanings of Addition and Subtraction

Unit 5

Strategies to Fluently Add within 100

Unit 6

Strategies to Fluently Subtract within 100

Unit 9
Strategies to Add 3-Digit Numbers

Unit 10
Strategies to Subtract 3-Digit Numbers

Unit 11

Data Analysis

Unit 12

Geometric Shapes and Equal Shares

Grade 2

Course Diagnostic

Name _____

Read each question carefully.

1. Look at the picture.

Which sentence matches the picture?

A. Ten and six more is 16.

B. Ten and seven more is 17.

C. Ten and eight more is 18.

D. Ten and nine more is 19.

2. What is the missing number?

80 − ? = 20

A. 30

B. 40

C. 50

D. 60

3. About how many blocks tall is the jack-in-the-box?

A. 3

B. 4

C. 6

D. 8

4. The circle is divided into equal parts.

How many equal parts are there?

A. 1

B. 2

C. 3

D. 4

5. Troy counts the number of dogs and cats that live in his neighborhood.

Dogs	卌 l
Cats	ll

How many more dogs than cats live in Troy's neighborhood?

A. 3

B. 4

C. 6

D. 8

Name _____

6. Look at the equations.

9 + 8 + 3 = 20

9 + ? = 20

What is the missing number?

_____

7. Look at the clock.

Which times are the same as the time shown on the clock? Choose all the correct answers.

A. 3 o'clock   **B.** half past 3:00

C. 2:30   **D.** half past 2:00

8. Which equation is true?

A. 5 − 2 = 8 − 5

B. 8 − 3 = 5 − 2

C. 5 − 3 = 8 − 3

D. 8 − 5 = 3 − 2

9. Look at the equations.

$$5 + ? = 13$$

$$13 - 5 = ?$$

Which number makes both equations true?

A. 7

B. 8

C. 13

D. 18

10. There are 6 apples, 8 oranges, and 4 bananas in a bowl.

How many pieces of fruit are in the bowl in all?

A. 12

B. 14

C. 16

D. 18

11. Look at the picture.

☆ ☆ ☆ ☆ ☆ ☆ ☆ ☆ ☆ ☆
☆ ☆ ☆ ☆ ☆ ☆ ☆ ☆ ☆ ☆
☆ ☆ ☆ ☆ ☆ ☆ ☆ ☆ ☆ ☆
☆ ☆ ☆ ☆ ☆ ☆ ☆ ☆ ☆ ☆
☆ ☆

How many stars are in the picture?

A. 24

B. 25

C. 42

D. 52

Name _____

12. What is the sum of 23 + 60? Use the base-ten blocks to find the sum.

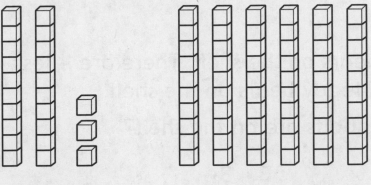

23 + 60 = _____

13. Which number is 10 less than 67?

A. 57

B. 66

C. 68

D. 77

14. Ahmed has 12 yogurts. He has 3 strawberry yogurts and the rest are blueberry.

How many of Ahmed's yogurts are blueberry?

A. 8

B. 9

C. 10

D. 12

15. What is the difference? Use the doubles fact to help you find the difference.

9 + 9 = 18

18 − 9 = _____

16. Does the shape have any circular bases?
Circle Yes or No for each shape.

Yes   No   Yes   No   Yes   No   Yes   No

17. There are 8 teddy bears on the shelf. There are 4 less stuffed rabbits than teddy bears on the shelf.

How many stuffed rabbits are on the shelf?

A. 2   **B.** 3

C. 4   **D.** 5

18. Dena makes a bookmark that is 10 inches long. Juan makes a bookmark that is 6 inches long. Abbie's bookmark is 3 inches longer than Juan's.

Who makes the longest bookmark?

A. Abbie

B. Dena

C. Juan

19. Which equation can be used to solve $8 + 7 = \square$?

A. $8 + 2 + 5 = \square$ **B.** $8 + 7 + 7 = \square$

C. $8 + 3 + 3 = \square$ **D.** $8 + 2 + 7 = \square$

20. What are the missing numbers?

82, 83, _____, 85, _____, 87

Lesson 1-1
Exit Ticket

Name _____

1. What did you learn about your teacher's math story?
   Write at least two things you learned.

2. What is one thing you want to become better at in math this year?

Reflect On Your Learning

Exit Ticket

Name _____

1. What strategies help you make sense of a problem?

2. What questions can you ask yourself to help you relate the numbers and quantities in a problem?

Reflect On Your Learning

Exit Ticket

Name _____

1. How can you use mathematics to describe a real-world problem?

2. Which tools can you use to solve addition and subtraction problems?

Reflect On Your Learning

Exit Ticket

Name _____

1. What are some strategies to explain your thinking to your classmates?

2. Why is it important to use correct terms when you are explaining your thinking?

Reflect On Your Learning

Lesson 1-5

Exit Ticket

Name _____

1. What are some patterns that you notice when adding and subtracting?

2. How can the patterns you notice help you add more efficiently?

Reflect On Your Learning

Exit Ticket

Name _____

1. What is your number one classroom norm for math class?

2. What is your personal goal for math this year?

Reflect On Your Learning

How Ready Am I?

Name _____

1. What number is the same as 10 ones?

   A. 00   **B.** 1

   C. 10   **D.** 100

2. What number is the same as 6 tens?

   A. 6   **B.** 16

   C. 60   **D.** 610

3. Which number is greater than 36?

   A. 30   **B.** 35

   C. 36   **D.** 37

4. Which of these is the same as 30?

   A. 1 + 1 + 1   **B.** 3 + 0

   C. 3 + 10   **D.** 10 + 20

5. What number is 10 more than 46?

   A. 56   **B.** 47

   C. 45   **D.** 36

6. Which comparison is true?

A. 51 > 15

B. 27 > 72

C. 63 < 36

D. 79 = 97

7. What is the value of the base-ten blocks?

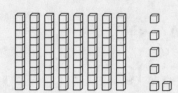

A. 68

B. 80

C. 86

D. 87

8. Caleb has 28 crayons. Josh has 10 fewer crayons than Caleb. How many crayons does Josh have?

A. 8

B. 18

C. 20

D. 28

9. Which of these is the same as 83?

A. 8 + 3

B. 8 + 30

C. 80 + 3

D. 80 + 30

10. Carlos has 49 toy cars. His brother has 37 toy cars. Which shows the correct comparison of the number of toy cars Carlos and his brother have?

A. 37 = 49

B. 37 > 49

C. 49 > 37

D. 49 < 37

Exit Ticket

Name _____

1. How many groups of 10 make 1 hundred? _____

2. Write the value of the base-ten blocks.

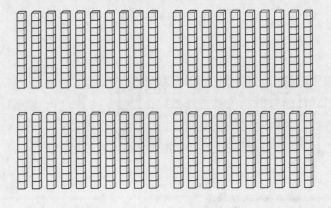

_____ tens = _____ hundreds = _____

3. How many groups of 10 make 300?

   A. 3   B. 10

   C. 30   D. 300

Reflect On Your Learning

Exit Ticket

Name _____

1. A 3-digit number has hundreds, _____, and ones.

2. What is the value of the 7 in the number 279?

A. 7   **B.** 10   **C.** 70   **D.** 6

3. Write the value of the base-ten blocks in the place-value chart.

hundreds	tens	ones

4. Which numbers have a 5 in the ones place? Choose all the correct answers.

A. 435   **B.** 530   **C.** 659   **D.** 985

Reflect On Your Learning

Exit Ticket

Name _____

1. You can read and write 3-digit numbers in standard form, expanded form, and _____ form.

2. Which of these does *not* represent the value of the blocks shown?

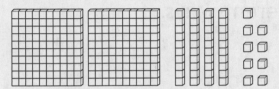

   A. 200 + 40 + 9   **B.** 249

   C. two hundred forty-nine   **D.** 2 + 4 + 9

3. What is six hundred fifty-nine in standard form?

   _____

4. There are 100 + 80 + 7 days until Caleb's birthday. What is the number of days in standard form?

   _____

Reflect On Your Learning

Exit Ticket

Name _____

1. Decompose 247 in two different ways.

   _____ + _____ + _____

   _____ + _____ + _____

2. Amie decomposes 845 in different ways.
   Complete Amie's work.

   8 hundreds, 4 tens, and _____ ones

   8 hundreds, 2 tens, and _____ ones

   8 hundreds, 1 ten, and _____ ones

   8 hundreds, 3 tens, and _____ ones

   8 hundreds, 0 tens, and _____ ones

3. Gabe has 186 trading cards. He wants to group
   the trading cards by tens and ones. How can he
   decompose 186 in two different ways using only
   tens and ones?

Reflect On Your Learning

Exit Ticket

Name _____

1. When comparing two 3-digit numbers, which place value do you compare first? _____

2. How can you compare the numbers? Complete with <, >, or =.

   246 ◯ 264

3. A zookeeper weighs a lion and a tiger. The lion weighs 418 pounds. The tiger weighs 481 pounds. Which comparisons of the animals' weights are correct? Choose all the correct answers.

   A. 418 = 481   **B.** 418 < 481

   C. 481 > 418   **D.** 481 < 418

Reflect On Your Learning

Performance Task

Name _____

Carnival Game Tickets

Liam and Samir play games at a carnival. They win tickets. They can trade in the tickets for prizes.

Part A

Liam plays Ring Toss and wins the tickets shown. How many points does he have?

_____ points

Part B

Liam plays Spin the Wheel and wins the tickets shown. How many of each ticket does he win? How many points?

_____ tens _____ ones

_____ points

Part C

Samir has 3 hundred-point tickets, 5 one-point tickets, and 7 ten-point tickets. How many total points is this?

Part D

Who has more points, Liam or Samir? Draw base-ten blocks or use a place-value chart to support your answer.

How do you know?

Part E

Samir trades one of his hundred-point tickets for tens and ones tickets. Write how many of each ticket he might have in two different ways.

_____ hundreds _____ tens _____ ones

_____ hundreds _____ tens _____ ones

Part F

Liam uses some of his tickets to get a toy shark. He then has 2 hundred-point tickets, I ten-point ticket, and 5 one-point tickets left. Does Liam have enough points left to get a bouncy ball that costs 100 points, a top that costs 20 points, and a ring that costs 5 points? Explain.

Unit Assessment, Form A

Name _____

I. Which shows 800?

    A. 800 tens   **B.** 80 ones

    C. 8 tens   **D.** 8 hundreds

2. Write the number in standard form.

two hundred thirty-seven _____

eight hundred thirteen _____

one hundred six _____

3. Which number do the base-ten blocks show?

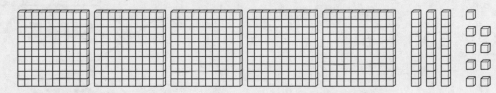

    A. 539   **B.** 593

    C. 935   **D.** 953

4. Match the number to its standard form.

6 tens, I one, 2 hundreds   126

I hundred, 6 ones, 2 tens   612

I ten, 2 ones, 6 hundreds   261

5. Decompose 536 in two different ways.

_____ + _____ + _____ = 536

_____ + _____ + _____ = 536

6. Which shows 418 in expanded form?

A. 400 + 10 + 8   **B.** 4 + 10 + 800

C. 400 + 1 + 8   **D.** 4 + 1 + 8

7. Peter is shopping for a cell phone, a tablet, and a camera. Match the item with the correct price tag. Not all price tags will be used.

Cell phone: 600 + 4 + 50

 $564

 $546

Tablet: 40 + 500 + 6

 $465

$456

Camera: 5 + 60 + 400 $654

8. Write <, >, or = to compare the pair of numbers.

396		401
219		319
180		108
771		717

Name _____

9. Which show 10 groups of ten?
   Choose all the correct answers.

   A. 100

   B.

   C.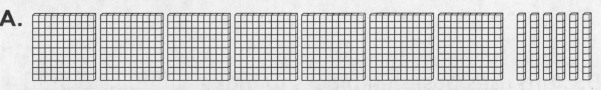

   D.

10. Julian does 10 sets of 10 sit-ups.
    How many sit-ups does Julian do in all?

    _____ _____

11. Which shows 627?

    A.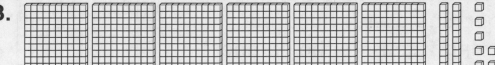

    B.

    C.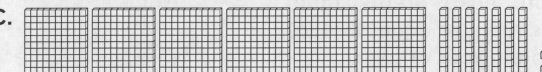

    D.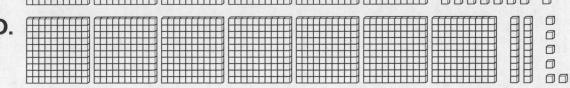

12. How can you show 764 in different ways? Write the missing numbers.

7 hundreds, 0 tens, and _____ ones

7 hundreds, 3 tens, and _____ ones

7 hundreds, 5 tens, and _____ ones

7 hundreds, 6 tens, and _____ ones

13. Which numbers are less than the number in the box? Choose all the correct answers.

$$\boxed{497}$$

A. 476

B. 498

C. 497

D. 490

14. How can you determine which of two 3-digit numbers is greater than the other? Explain.

15. Clara uses base-ten blocks to make a number. She uses 4 hundreds, 8 ones, and 2 tens. Is the number Clara makes greater than, less than, or equal to 436? Explain your answer.

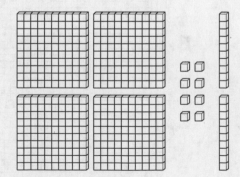

Unit 2
Unit Assessment, Form B

Name _____

1. Which shows 600?

   A. 600 tens   **B.** 6 hundreds

   C. 6 tens   **D.** 60 ones

2. Write the number in standard form.

   six hundred twenty-one _____

   two hundred eighty _____

   nine hundred one _____

3. Which number do the base-ten blocks show?

   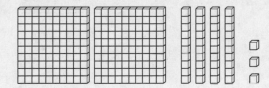

   A. 234   **B.** 243

   C. 423   **D.** 432

4. Match the number to its standard form.

   1 ten, 4 hundreds, and 6 ones   146

   6 hundreds, 1 one, and 4 tens   416

   4 tens, 6 ones, and 1 hundred   641

5. Decompose 356 in two different ways.

_____ + _____ + _____ = 356

_____ + _____ + _____ = 356

6. Which shows 865 in expanded form?

A. 800 + 6 + 5   **B.** 8 + 6 + 5

C. 8 + 60 + 500   **D.** 800 + 60 + 5

7. Micah's family is shopping for a stove, a refrigerator, and a dishwasher. Match the item with the correct price tag. Not all price tags will be used.

Stove: 400 + 2 + 70

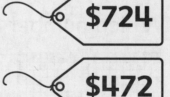

Refrigerator: 20 + 4 + 700

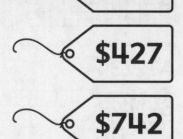

Dishwasher: 7 + 200 + 40

8. Write <, >, or = to compare the pair of numbers.

443		891
201		193
529		592
764		746

9. Which show 10 groups of ten?
Choose all the correct answers.

A. 100

B.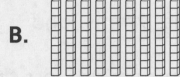

C.

D.

10. Maria does 10 sets of 20 jumping jacks.
How many jumping jacks does Maria do in all?

_____ _____

11. Which shows 546?

A.

B.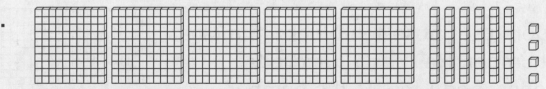

C.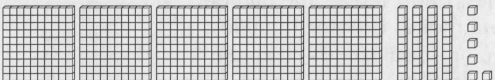

D.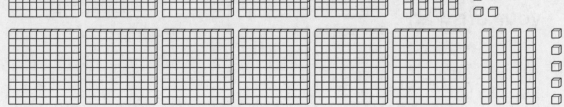

12. How can you show 598 in different ways? Write the missing numbers.

5 hundreds, 0 tens, and _____ ones

5 hundreds, 5 tens, and _____ ones

5 hundreds, 8 tens, and _____ ones

5 hundreds, 9 tens, and _____ ones

13. Which numbers are less than the number in the box? Choose all the correct answers.

$$\boxed{823}$$

A. 832   **B.** 813

C. 822   **D.** 830

14. How can you determine which of two 3-digit numbers is greater than the other? Explain.

15. Artem uses base-ten blocks to make a number. He uses 3 hundreds, 4 ones, and I ten. Is the number Artem makes greater than, less than, or equal to 304? Explain your answer.

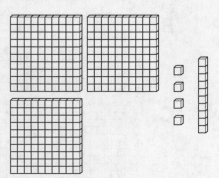

How Ready Am I?

Name _____

1. Which numbers add to make 7?

   A. 2 and 5   **B.** 1 and 8

   C. 4 and 2   **D.** 5 and 3

2. What is the sum of 5 + 5?

   A. 7   **B.** 8   **C.** 9   **D.** 10

3. Which number is 10 more than 73?

   A. 70   **B.** 74   **C.** 80   **D.** 83

4. Count by 1s. What is the next number?

   54, 55, 56, 57, 58, __

   A. 58   **B.** 59   **C.** 60   **D.** 68

5. Karina has 5 roses. Shelby has 4 roses. Which equation shows how many roses they have in all?

   A. 9 − 5 = 4   **B.** 5 + 4 = 9

   C. 5 − 4 = 1   **D.** 9 + 4 = 13

6. Which doubles fact equals 8?

    A. 2 + 2   **B.** 4 + 4

    C. 5 + 5   **D.** 8 + 8

7. Which represents 17?

    A. 1 + 7   **B.** 10 − 7

    C. 1 − 7   **D.** 10 + 7

8. Count by 10s. What is the next number?

50, 60, 70, 80, ___

    A. 81   **B.** 85   **C.** 90   **D.** 100

9. Look at the blocks.
Which number shows how many?

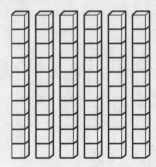

    A. 50   **B.** 60

    C. 66   **D.** 70

10. Which number do the counters in the array represent?

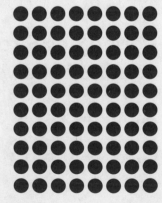

    A. 64   **B.** 72

    C. 80   **D.** 100

Exit Ticket

Name _____

1. What are the missing numbers?

423		425	426			429	430		432
433	434		436	437	438		440	441	
	444	445		447	448	449	450	451	452

2. Riley is counting by 1s. What are the missing numbers?

   356, _____, _____, _____, _____, _____, 362

3. Heidi counts 458, 459, 460, 461. Then she counts the next five numbers. Which of these are true about the next five numbers? Choose all the correct answers.

   A. They all have a 4 in the hundreds place.

   B. They all have a 1 in the ones place.

   C. They all have a 5 in the tens place.

   D. They all have a 6 in the tens place.

Reflect On Your Learning

Exit Ticket

Name _____

1. Start at 708. Count by 5s until you reach 738.
   Complete the sentence to describe the pattern.

   Every number ends with _____ or _____.

2. Start at 611. Count by 5s. Which of these numbers do
   you count? Choose all the correct answers.

   A. 615   **B.** 616

   C. 630   **D.** 636

3. A teacher has 124 markers. She buys 7 more boxes of
   markers. There are 5 markers in each box. How many
   markers does the teacher have now? Explain.

Reflect On Your Learning

Exit Ticket

Name _____

1. Start at 0. Count by 100s.
   Which of these describes the pattern?

   A. The digit in the hundreds place goes up by 1. The digits in the tens place and the ones place stay the same.

   B. The digits in the hundreds place and the tens place go up by 1. The digit in the ones place stays the same.

   C. The digits in the hundreds place and the tens place stay the same. The digit in the ones place goes up by 1.

2. Start at 227. Skip count by 10s until you reach 297. What is the pattern?

3. Gina has 40 trading cards. She gets 3 more packs. Each pack has 10 trading cards. Gina says she now has 80 trading cards. How do you respond to her?

Reflect On Your Learning

Exit Ticket

Name _____

1. Piper sees an even number of butterflies. Which number of butterflies could be the number Piper sees? Choose all the correct answers.

   A. 11 butterflies   **B.** 14 butterflies

   C. 15 butterflies   **D.** 18 butterflies

2. There is an odd number of fish in a fish tank. There are more than 15 fish and less than 20 fish. How many fish could be in the tank? Choose all the correct answers.

   A. 15 fish   **B.** 16 fish

   C. 17 fish   **D.** 19 fish

3. A teacher tells 23 students to pair up with a reading buddy. Will each student have a reading buddy? Explain.

Reflect On Your Learning

Exit Ticket

Name _____

1. _____ numbers can be separated into two equal groups.

2. Which equations have an even sum?
   Choose all the correct answers.

   A. 2 + 3 = ?   **B.** 4 + 4 = ?

   C. 7 + 7 = ?   **D.** 8 + 9 = ?

3. A vase has 6 red tulips and 7 yellow tulips.
   Is the total number of tulips in the vase odd or even?
   Explain how you know.

Reflect On Your Learning

Exit Ticket

Name _____

1. How can you skip count to find the number of counters in the array? Choose all the correct answers.

   A.   4, 8, 12, 16

   B.   5, 10, 15, 20

   C.   4, 8, 12, 16, 20

   D.   5, 10, 15, 20, 25

2. How many paper clips are in the array? Skip count.

   _____ paper clips

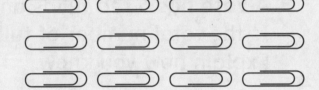

3. Will says there is only one way he can arrange 9 trading cards in an array that has more than one row and more than one column. How do you respond to him? Explain.

Reflect On Your Learning

Exit Ticket

Name _____

1. Write two equations that match the array.

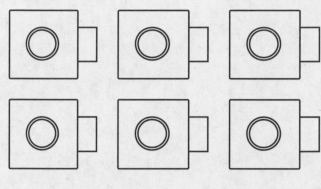

___ + ___ = ___   ___ + ___ + ___ = ___

2. Draw *all* of the arrays that match the equation
4 + 4 + 4 = 12.

3. Sophia's puzzle has 3 rows of 5 pieces. Miguel's puzzle has 5 rows of 3 pieces. Whose puzzle has more pieces?

Reflect On Your Learning

Performance Task

Name _____

Party Planning

Beth is having a party.

Part A

Beth needs marbles for a game she is playing at the party.
The marbles come in packages of 100.
Beth buys 7 packages of marbles.

Show how Beth can count the marbles she buys.

100, 200, _____, _____, _____, _____,

How many marbles does she buy altogether?

Part B

Beth also needs to buy plates.
Beth buys these packages of plates.

How many plates does she buy?

Part C

Beth buys 6 green balloons and 7 pink balloons.

Explain how you know Beth has an odd number of balloons.

How could Beth get to an even number of balloons?

Part D

Beth sets up chairs for the magic show she is having at her party. She sets up 3 rows of 5 chairs.

Draw an array to show the chairs.

Write an equation to describe the number of chairs Beth sets up.

How many chairs does Beth set up?

Unit Assessment, Form A

Name _____

I. Mario is counting by Is. He counts 266, 267, 268. What are the next five numbers that he counts?

_____, _____, _____, _____, _____

2. Count by 5s to determine the missing numbers in the pattern. What are the missing numbers?

_____, 125, 130, _____, 140, _____

3. Maya is reading a book.

- She reads 5 pages on Monday.

- She reads 10 pages on Tuesday.

- She reads 15 pages on Wednesday.

If Maya keeps reading with this pattern, how many pages does she read on Thursday?

A. 15 pages   **B.** 20 pages

C. 25 pages   **D.** 30 pages

4. How can you skip count to find the number of counters in the array? Choose all the correct answers.

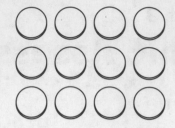

    A. 3, 6, 9

    B. 4, 8, 12

    C. 3, 6, 9, 12

    D. 4, 8, 12, 16

5. Which array shows $5 + 5 + 5$?

    A.
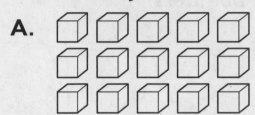

    B.
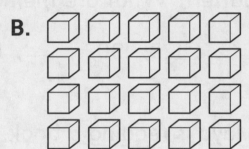

    C.

    D.
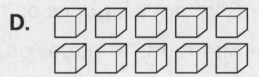

6. A store displays 18 cups of yogurt on 3 shelves. Which of these is the correct equation for finding the total number of yogurt cups?

    A. $6 + 3 = 18$ **B.** $6 + 6 + 6 = 18$

    C. $3 + 3 + 3 = 18$ **D.** $6 + 6 + 6 + 6 + 6 + 6 = 18$

Name _____

7. Skip count by 10s.
   Write the numbers to complete the number line.

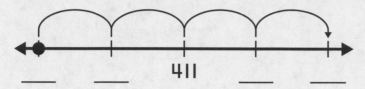

_____ _____   411   _____ _____

8. Start at 118. Count by 100s. Which of these numbers
   do you count? Choose all the correct answers.

   A. 200   **B.** 208

   C. 218   **D.** 318

9. Which equations have an even sum?
   Choose all the correct answers.

   A. $5 + 5 = ?$ **B.** $6 + 7 = ?$

   C. $8 + 9 = ?$ **D.** $9 + 9 = ?$

10. Mel is having blueberries for
    a snack. Is the number of
    blueberries even or odd?
    Explain.

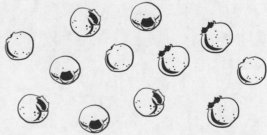

11. Rolanda has 10 hair bows. Draw two different arrays to show two ways Rolanda can arrange the bows.

12. Gavin finds some shells on the beach. Does Gavin find an even number of shells? Explain.

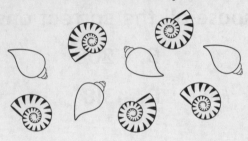

Unit Assessment, Form B

Name _____

1. Leon is counting by Is. He counts 514, 515, 516. What are the next five numbers that he counts?

   ____, ____, ____, ____, ____

2. Count by 5s to determine the missing numbers in the pattern. What are the missing numbers?

   ____, 170, 175, ____, 185, ____

3. Marcus is practicing the violin.

   • He practices for 35 minutes on Monday.

   • He practices for 40 minutes on Tuesday.

   • He practices for 45 minutes on Wednesday.

   If Marcus keeps practicing the violin with this pattern, how many minutes does he practice on Thursday?

   A. 45 minutes   **B.** 50 minutes

   C. 51 minutes   **D.** 55 minutes

4. How can you skip count to find the number of counters in the array? Choose all the correct answers.

   A. 3, 6, 9

   B. 5, 10, 15

   C. 3, 6, 9, 12, 15

   D. 5, 10, 15, 20, 25

5. Which array shows $4 + 4 + 4 + 4 + 4$?

   A. 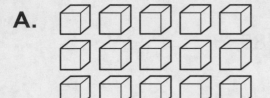   **B.**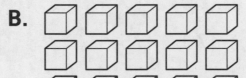

   C.    **D.**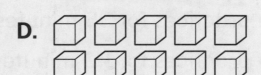

6. Mrs. Jones uses 15 lemons to make lemonade. Which of these is the correct equation for finding the total number of lemons?

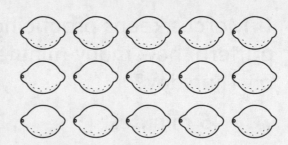

   A. $5 + 3 = 15$ **B.** $3 + 3 + 3 = 15$

   C. $5 + 5 + 5 = 15$ **D.** $5 + 5 + 5 + 5 + 5 = 15$

Name

7. Skip count by 10s.
Write the numbers to complete the number line.

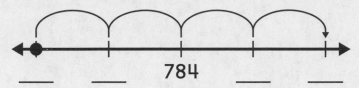

____ ____ 784 ____ ____

8. Start at 265. Count by 100s. Which of these numbers do you count? Choose all the correct answers.

A. 300   **B.** 365

C. 400   **D.** 465

9. Which equations have an odd sum?
Choose all the correct answers.

A. $5 + 5 = ?$ **B.** $6 + 7 = ?$

C. $8 + 9 = ?$ **D.** $9 + 9 = ?$

10. Nikhil puts blueberries in his yogurt. Is the number of blueberries even or odd? Explain.

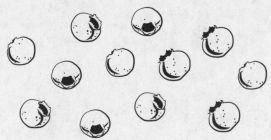

11. Hadley has 8 photos. Draw two different arrays to show two ways Hadley can arrange the photos.

12. Beatriz collects some shells on the shore. Does she collect an even number of shells? Explain.

How Ready Am I?

Name _____

1. What is the unknown number?
   $3 + 5 = \square$

   A. 2   **B.** 7   **C.** 8   **D.** 9

2. Which of these can help you find the unknown
   number in $7 - \square = 4$?

   A. $4 + 3$ **B.** $5 + 2$ **C.** $4 + 7$ **D.** $7 + 4$

3. What is the unknown number?
   $12 - \square = 8$

   A. 2   **B.** 4   **C.** 6   **D.** 8

4. Lorita has 6 oranges. She buys 4 more oranges. Which
   of these can help you find how many oranges Lorita
   has in all?

   A. $6 - 4$ **B.** $6 + 4$ **C.** $10 - 4$ **D.** $10 + 4$

5. What is the sum of $6 + 5$?

   A. 11   **B.** 12   **C.** 13   **D.** 14

6. Scott solves a problem by counting "8, 9, 10, 11."

Which of these could be the problem Scott solves?

A. There are 7 apples on a tree. 4 apples fall. How many apples are left?

B. There are 5 cats. There are 9 dogs. How many cats and dogs are there in all?

C. There are 7 frogs in a pond. 4 frogs hop away. How many frogs are left?

D. There are 7 birds on a branch. 4 more birds join them. How many birds are there now?

7. Kim picks 13 flowers. She gives some to a friend. She has 8 flowers left. Use the equation to help you solve the problem.

$13 - ? = 8$

How many flowers does Kim give to a friend?

A. 5   **B.** 6   **C.** 7   **D.** 8

8. Nhi has 12 grapes. She gives 4 away. Which of these can help you find how many grapes Nhi has now?

A. $8 - 4$ **B.** $12 + 4$ **C.** $12 - 4$ **D.** $16 + 4$

9. Which of these can help you find the missing number in $8 + \square = 13$?

A. $8 - 5$ **B.** $8 - 13$ **C.** $13 - 6$ **D.** $13 - 8$

10. Which equation represents the word problem? Lidia has 14 coins. She gives 8 coins away. How many coins does Lidia have now?

A. $14 - 6 = ?$ **B.** $14 - 8 = ?$ **C.** $14 + 8 = ?$ **D.** $14 + 6 = ?$

Exit Ticket

Name _____

1. Which equation represents the word problem?

   Henry sees some starfish on the beach. Then he sees 9 more starfish in the water. Henry sees 28 starfish in all. How many starfish does Henry see on the beach?

   A. $37 - ? = 28$

   B. $9 + 28 = ?$

   C. $37 - 9 = ?$

   D. $? + 9 = 28$

2. Write an equation to represent the problem using ? for the unknown. Then solve.

   Lucy plays a board game. She earns some points. Then Lucy earns 7 more points. Now Lucy has 61 points. How many points does Lucy have before?

   Equation: _____

   Solve: _____

Reflect On Your Learning

Exit Ticket

Name _____

1. Match the equation to the word problem it represents.

There are some ducks in a
pond. 7 ducks fly away. There
are 23 ducks left. How many
ducks were in the pond before?

$23 - ? = 7$

There are 23 children at a park.
Some children leave. 7 children
are still at the park. How many
children leave?

$23 - 7 = ?$

Tess has 23 carrot sticks. She
eats 7 carrot sticks. How many
carrot sticks are left?

$? - 7 = 23$

2. Write an equation to represent the problem using ?
for the unknown. Then solve.

Kim has 11 lemons. She uses some lemons to make
lemonade. There are 7 lemons left. How many lemons
does Kim use to make lemonade?

Equation: _____

Solve: _____ _____

Reflect On Your Learning

Exit Ticket

Name _____

1. Which steps represent how to solve the word problem?

   There are 8 bees on the flowers. 3 bees fly away. 6 more bees land on the flowers. How many bees are on the flowers now?

   A. Add 8 and 3. Then add 6 to the sum.

   B. Add 8 and 6. Then add 3 to the sum.

   C. Subtract 3 from 8. Then add 6 to the difference.

   D. Subtract 6 from 8. Then add 3 to the difference.

2. Complete the equation to represent the problem. Then solve.

   Jin has 12 blueberries. She eats 4 blueberries. Her mother gives her 7 blueberries. How many blueberries does Jin have now?

   Equation: 12 _____ 4 _____ 7 = ?

   Solve: _____ _____

Reflect On Your Learning

Exit Ticket

Name _____

1. Complete the part-part-whole mat to represent the problem. Use ? for the unknown.

   There are 31 books on Clay's bookshelf. 8 books are about sports and the rest are about animals. How many animal books are on Clay's bookshelf?

Part	Part
Whole	

2. Write two equations to represent the problem using ? for the unknown. Then solve.

   Amelia has 13 dolls. 4 of the dolls have blonde hair. The rest of the dolls have brown hair. How many dolls have brown hair?

   Equations:

   Solve: _____ _____

Reflect On Your Learning

Exit Ticket

Name _____

1. Which equations represent the word problem? Choose all the correct answers.

   James is making a fruit salad. He uses 42 berries. 4 of the berries are strawberries and the rest are blackberries. How many blackberries does James use?

   A. $42 + 4 = ?$

   B. $42 - 4 = ?$

   C. $4 + ? = 42$

   D. $? - 42 = 4$

2. Carmen makes 16 bracelets. 11 bracelets are pink and the rest are red. Carmen writes the equation $16 - 11 = ?$ to find the number of red bracelets. Write another equation that Carmen can use to find the number of red bracelets.

Reflect On Your Learning

Exit Ticket

Name _____

1. Write an equation to represent the problem using ? for the unknown.

There are 10 seals swimming in the water. 5 seals get out of the water. Then some seals get in the water. Now there are 8 seals in the water. How many seals get in the water?

2. Complete the equation to represent the problem. Then solve.

Gilda has 9 flowers. She picks some more flowers. Then she gives 8 flowers to her mother. Gilda has 5 flowers left. How many flowers does Gilda pick?

Equation: _____ + _____ − _____ = 5

Solve: _____ _____

Reflect On Your Learning

Lesson 4-7
Exit Ticket

Name _____

1. Which equations represent the word problem?
Choose all the correct answers.

Martin folds 45 paper airplanes. Martin folds 7 fewer paper airplanes than Pilar. How many paper airplanes does Pilar fold?

A. $45 - 7 = ?$

B. $? - 45 = 7$

C. $45 + 7 = ?$

D. $7 + ? = 45$

2. Write an equation to represent the problem using ? for the unknown. Then solve.

Pablo buys 12 onions. He buys 6 fewer onions than potatoes. How many potatoes does Pablo buy?

Equation:

Solve: _____ _____

Reflect On Your Learning

Exit Ticket

Name _____

1. Ebony runs 12 laps. She runs 3 more laps than Donald. The equation $12 - 3 = ?$ can be used to find the number of laps Donald runs.

   Write another equation that can be used to find the number of laps Donald runs.

2. Write an equation to represent the problem using ? for the unknown. Then solve.

   Heidi collects 25 eggs from her chickens. She collects 8 more eggs than Cory. How many eggs does Cory collect?

   Equation:

   Solve: _____

Reflect On Your Learning

Exit Ticket

Name _____

1. Complete the diagram to represent the problem.

Olga does 15 jumping jacks. Dane does 6 more jumping jacks than Olga. How many jumping jacks do Olga and Dane do in all?

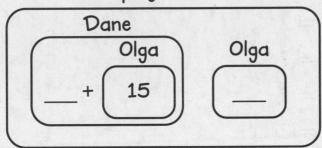

? jumping jacks in all

Dane

Olga   Olga

____ + [15]   ____

_____ _____

2. Write an equation to represent the problem using ? for the unknown. Then solve.

Polly uses 9 wooden pieces to make a train track. Lane uses 2 more wooden pieces than Polly to make a train track. How many wooden pieces do Polly and Lane use in all?

Equation:

Solve: _____ _____

Reflect On Your Learning

Exit Ticket

Name _____

1. Which equation represents the word problem? Choose the correct answer.

   There are 9 ladybugs in the grass. 5 ladybugs fly away. 3 more ladybugs land in the grass. How many ladybugs are in the grass now?

   A. $9 + 5 + 3 = ?$

   B. $9 - 5 + 3 = ?$

   C. $9 + 5 - 3 = ?$

   D. $9 - 5 - 3 = ?$

2. How can you solve the problem? Write the steps. Then solve.

   There are 13 people at the beach. 5 more people come to the beach. Then 4 people leave the beach. How many people are at the beach now?

   Solve: _____ _____

Reflect On Your Learning

Performance Task

Name _____

Soccer

Kaden and Jenna are playing soccer with friends at recess.

Part A

The game starts with 15 players. Then 7 more players join. How many players are there now? Write an equation to represent the problem. Then solve the problem.

Part B

Jenna has 12 players on her team. 4 players play defense. The rest play offense. How many players play offense? Complete the chart to represent the problem. Write an equation. Then solve the problem.

Part	Part
Whole	

Part C

Kaden's team scores 6 more goals than Jenna's team. Jenna's team scores 9 goals. How many goals does Kaden's team score? Write two different equations to represent the problem. Then solve the problem.

Part D

Kaden scores 7 goals. Some are in the first half. The rest are in the second half. How many goals could Kaden have scored in each half? Write three possible combinations.

Part E

Jenna plays 6 fewer minutes in the first half than she plays in the second half. Jenna plays 13 minutes in the second half. How many minutes does Jenna play in all? Explain how you solved the problem.

Unit Assessment, Form A

Name _____

1. Match the equation to the word problem it represents.

Carl has 11 math problems for homework. He solves 4 math problems. How many math problems are left?
$11 - ? = 4$

There are 11 clowns in a parade. Some clowns are juggling. 4 clowns are dancing. How many clowns are juggling?
$11 - 4 = ?$

There are some squirrels in a tree. 4 squirrels leave. There are 11 squirrels left. How many squirrels were in the tree before?
$? - 4 = 11$

2. Which equation represents the word problem?

Walter has some blue toy cars. He also has 2 red toy cars. He has 16 red and blue cars in all. How many blue cars does Walter have?

A. $? + 2 = 16$

B. $2 + 16 = ?$

C. $18 - ? = 16$

D. $18 - 2 = ?$

3. Write an equation to represent the problem using ? for the unknown. Then solve.

Jesse puts 8 animal stickers in his sticker book. He puts 3 more truck stickers than animal stickers in his sticker book. How many animal and truck stickers does Jesse put in his sticker book in all?

Equation:

Solve: _____ _____

4. Write an equation to represent the problem.

Julia has 8 stuffed toys. She gives 5 to her sister. Then she gets some stuffed toys for her birthday. Now she has 10 stuffed toys. How many stuffed toys did she get for her birthday?

5. How can you solve the problem? Choose the correct answer.

Marco's mom gives him 19 grapes. He eats 8 grapes. His mom gives him 4 more grapes. How many grapes does Marco have now?

A. Add 8 and 4. Then add 19 to the sum.

B. Add 19 and 8. Then add 4 to the sum.

C. Subtract 4 from 8. Then add 19 to the difference.

D. Subtract 8 from 19. Then add 4 to the difference.

6. Which equation represents the word problem? Choose the correct answer.

   Kyle rode his bike on Saturday and Sunday. He rode his bike for 8 minutes on Saturday. He rode his bike a total of 59 minutes on both days. How many minutes did Kyle ride his bike on Sunday?

   A. $8 + 59 = ?$ **B.** $67 - 59 = ?$

   C. $67 - ? = 8$ **D.** $? + 8 = 59$

7. Al has 10 blue shirts. He has 4 more blue shirts than green shirts. The equation $10 - 4 = ?$ can be used to find the number of green shirts Al has. Write another equation that can be used to find the number of green shirts Al has.

8. Complete the equation to represent the problem. Then solve.

   Kevin makes 13 snowballs. He throws 7 snowballs. He makes 5 more snowballs. How many snowballs does Kevin have now?

   Equation: 13 _____ 7 _____ 5 = ?

   Solve: _____ _____

9. Write two equations to represent the problem using ? for the unknown. Then solve.

Justin sees 24 birds while walking at a park. 8 of the birds are robins. The rest are sparrows. How many sparrows does Justin see?

Equations:

Solve: _____ _____

10. Morgan writes 15 thank-you notes. Sara writes 4 fewer thank-you notes than Morgan. Can the equation 15 + 4 = ? be used to find how many thank-you notes Sara writes? Explain.

11. What equation can represent the problem? Solve and explain how your equation relates to the problem.

Andy has 30 new e-mails to read. He reads 6 e-mails. How many e-mails does he have left to read?

Unit Assessment, Form B

Name _____

I. Match the equation to the word problem it represents.

Lina needs to practice the piano for 15 minutes. She practices for 6 minutes. How many minutes are left?

$$? - 6 = 15$$

There are 15 floats in a parade. Some floats have people on them. 6 floats do not. How many floats have people on them?

$$15 - ? = 6$$

There are some leaves in a tree. 6 leaves drop to the ground. There are 15 leaves left. How many leaves were in the tree before?

$$15 - 6 = ?$$

2. Which equation represents the word problem?

Some balls are in a bin. Luca puts 5 more balls in the bin. Now there are 21 balls in the bin. How many balls were in the bin before?

A. $5 + 21 = ?$ **B.** $? + 5 = 21$

C. $26 - 5 = ?$ **D.** $26 - ? = 21$

3. Write an equation to represent the problem using ? for the unknown. Then solve.

Clara makes 7 bookmarks. Diego makes 4 more bookmarks than Clara. How many bookmarks do Clara and Diego make in all?

Equation:

Solve: _____ _____

4. Write an equation to represent the problem.

Mark has 28 pictures of his dog on his phone. He deletes 5 of the pictures. He takes some more pictures of his dog. Now there are 29 pictures of his dog on his phone. How many more pictures did Mark take?

5. How can you solve the problem? Choose the correct answer.

Ben's mom gives him 7 crackers. He eats 6 crackers. His mom gives him 3 more crackers. How many crackers does Ben have now?

A. Add 7 and 6. Then add 3 to the sum.

B. Add 7 and 3. Then add 6 to the sum.

C. Subtract 6 from 7. Then add 3 to the difference.

D. Subtract 3 from 7. Then add 6 to the difference.

Name _____

6. Which equation represents the word problem? Choose the correct answer.

There are 68 tacos for sale at a taco stand. People buy some tacos. Now 10 tacos are left at the taco stand. How many tacos do people buy?

A. $68 + 10 = ?$ 　　　　　 B. $78 - ? = 10$

C. $10 + ? = 68$ 　　　　　 D. $78 - 68 = ?$

7. Hugo draws 6 pictures. He draws 2 more pictures than Janna. The equation $6 - 2 = ?$ can be used to find the number of pictures Janna draws. Write another equation that can be used to find the number of pictures Janna draws.

8. Complete the equation to represent the problem. Then solve.

Mila has 12 stickers. She gives 4 stickers to her sister. She buys 6 stickers. How many stickers does Mila have now?

Equation: 12 _____ 4 _____ 6 = ?

Solve: _____

9. Write two equations to represent the problem using ? for the unknown. Then solve.

Theo played sports outside for 15 minutes. He practiced playing catch for 7 minutes. He spent the rest of the time playing basketball. How many minutes did Theo play basketball?

Equations:

Solve: _____ _____

10. Marcel hangs 14 pictures of his family on a wall. He hangs 8 fewer pictures of his pets than of his family. Can the equation $14 + 8 = ?$ be used to find how many pictures of his pets Marcel hangs?

11. What equation can represent the problem? Solve and explain how your equation relates to the problem.

Hannah has 17 tulip bulbs to plant in her garden. She plants 9 tulip bulbs. How many tulip bulbs does she have left to plant?

Benchmark Assessment 1

Name _____

1. Bo has 6 blocks. Liz has 9 more blocks than Bo. How many blocks does Liz have?

 A. 3 blocks

 B. 5 blocks

 C. 13 blocks

 D. 15 blocks

2. Dora has an odd number of stickers. Which number could be the amount of stickers Dora has?

 A. 8

 B. 11

 C. 12

 D. 16

3. Which comparison is true?

 A. 374 < 376

 B. 367 > 376

 C. 367 > 374

 D. 376 < 364

4. Which picture represents 100?

A.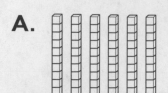

B.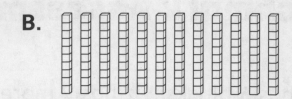

C.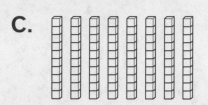

D.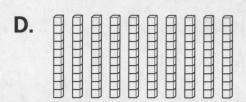

5. Which set of numbers shows skip counting by 5s?

    A. 500, 600, 700

    B. 710, 715, 720

    C. 350, 450, 550

    D. 910, 920, 930

6. Julie buys a salad with 6 tomatoes on it. She puts 7 more tomatoes on the salad.

How many tomatoes are on the salad in all?

_____ tomatoes

7. Pilar counts by 1s starting from 617.
Fill in the next three numbers that Pilar counts.

617, 618, 619, _____, _____, _____

8. Davis has 14 peanuts. He gives some to a friend.

   Davis has 8 peanuts left.

   How many peanuts does Davis give to his friend?

   14 − ? = 8

   _____ peanuts

9. Which group has an even number of circles?

   A.

   B.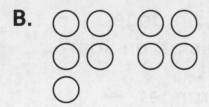

   C.

   D.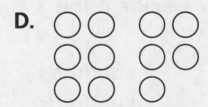

10. Look at the picture.

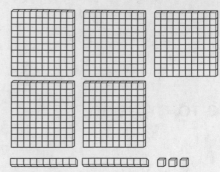

Which number does the picture show? Choose all the correct answers.

A. five thousand twenty-three

B. 523

C. five hundred twenty-three

D. 5,203

E. 500 + 20 + 3

11. Lucy picked 11 oranges. Sam picked 3 oranges. How many more oranges did Lucy pick than Sam?

A. 8 oranges   **B.** 9 oranges

C. 11 oranges   **D.** 14 oranges

12. What do the digits in 628 represent? Decide if the digit shows ones, tens, or hundreds.

6	ones	tens	hundreds
2	ones	tens	hundreds
8	ones	tens	hundreds

Name _____

13. Is the number equal to 630? Choose Yes or No.

	Yes	No
6 + 3 + 0		
six hundred thirty		
600 + 30		
six hundred three		

14. Look at the array of marbles.

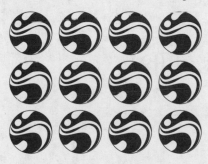

a. Which expressions match the array? Choose all the correct answers.

A. 3 + 3 + 3 + 3

B. 3 + 3 + 3

C. 4 + 4 + 4

D. 4 + 4 + 4 + 4

E. 3 + 3 + 3 + 3 + 3

b. How many marbles are in the array?

_____ marbles

15. Guess the number.

- I am less than 3 hundreds, 5 tens, and 2 ones.

- I am greater than 3 hundreds and 5 tens.

What number am I?

_____

16. Dante has a package of 12 yogurts. 5 are blueberry and the rest are strawberry. He eats 3 strawberry yogurts.

How many strawberry yogurts does Dante have left?

_____ strawberry yogurts

17. Jayden skip counts by 100s. He starts at 119.

119, 219, 319, _____

Which of these is the missing number?

A. 320   **B.** 329

C. 419   **D.** 420

18. Yuri has 4 pears. Then she buys 10 more pears. How many pears does Yuri have now?

_____ pears

How Ready Am I?

Name _____

1. Malika has 3 pencils. Her brother gives her 7 pencils. Then she buys 6 more pencils. How many pencils does Malika have now?

   A. 10 pencils   **B.** 13 pencils

   C. 16 pencils   **D.** 18 pencils

2. What is the sum? Use the base-ten blocks to add.

   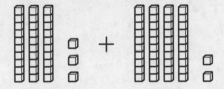

   A. 12   **B.** 42   **C.** 57   **D.** 75

3. What is the missing number?
   7 + ? = 10

   A. 2   **B.** 4   **C.** 3   **D.** 1

4. On Monday, Maya reads 11 pages of a book. Today, she reads 7 more pages. How many pages of the book does Maya read in all?

   A. 22 pages   **B.** 18 pages

   C. 16 pages   **D.** 12 pages

5. Rachel makes 9 baskets at practice. When she gets home, she makes 7 more baskets. How many baskets does Rachel make in all? Use the number line to help you.

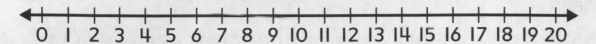

0 1 2 3 4 5 6 7 8 9 10 11 12 13 14 15 16 17 18 19 20

A. 12 baskets   **B.** 15 baskets

C. 16 baskets   **D.** 18 baskets

6. Which shows a correct way to make 52?

A. 50 + 20   **B.** 50 + 2

C. 20 + 5   **D.** 5 + 2

7. What is the sum of 45 + 30?

A. 75   **B.** 70   **C.** 65   **D.** 48

8. What is the sum of 8 + 7 + 3?

A. 18   **B.** 15   **C.** 11   **D.** 10

9. What number do the base-ten blocks show?

A. 14   **B.** 20

C. 24   **D.** 42

10. Which doubles fact is *not* correct?

A. 5 + 5 = 10   **B.** 6 + 6 = 12

C. 7 + 7 = 13   **D.** 8 + 8 = 16

Exit Ticket

Name _____

1. Sue is solving $8 + 7 = ?$. How can she decompose the second addend to make a 10 with the first addend?

   A. $0 + 7$ **B.** $1 + 6$

   C. $2 + 5$ **D.** $3 + 4$

2. Kya uses the number line shown to find the sum of two numbers. Which number sentence matches Kya's work?

   A. $7 + 6 = 13$ **B.** $8 + 5 = 13$

   C. $8 + 5 = 14$ **D.** $9 + 4 = 13$

3. There are 7 red birds and 6 blue birds in a tree. How many birds are in the tree in all? Make a 10 to help you find the answer.

Reflect On Your Learning

Exit Ticket

Name _____

1. Neill says he can find the sum of 6 + 8 by making a double and adding 2 more. Which doubles fact does Neill make?

   A. 3 + 3   **B.** 4 + 4

   C. 5 + 5   **D.** 6 + 6

2. What is the sum of 7 + 8? Solve using a doubles fact.

   A. 14   **B.** 15

   C. 16   **D.** 17

3. Remy plays 9 games of cards and 9 games of dominoes. How many games does Remy play? Write the doubles fact that solves the problem.

Reflect On Your Learning

Lesson 5-3
Exit Ticket

Name _____

1. Which expressions need to use regrouping to add? Choose all the correct answers.

   A. 20 + 59   B. 42 + 28

   C. 35 + 55   D. 62 + 17

2. There are 28 people in a movie theater. Then, 37 more people walk into the theater. How many people are in the theater now? Use base-ten shorthand to solve the problem.

3. Tori says 39 + 37 = 616. What error did Tori make?

   A. Tori did not count all the ones when she added 9 ones to 7 ones.

   B. Tori did not regroup 10 ones as 1 ten.

   C. Tori added 3 tens to 7 tens instead of adding 3 tens to 3 tens.

   D. Tori did not count all the tens when she added 3 tens to 3 tens.

Reflect On Your Learning

Exit Ticket

Name _____

1. Which equation has the same sum as 14 + 21?

   A. 41 + 21 = ?   B. 14 + 12 = ?

   C. 21 + 14 = ?   D. 21 + 41 = ?

2. Match the equations that have the same sum.

   15 + 34 = ?   48 + 27 = ?

   27 + 48 = ?   43 + 51 = ?

   51 + 43 = ?   34 + 15 = ?

3. What is the sum?

   19 + 57 = _____

4. There are 28 girls and 32 boys that ride a school bus.
   Which equations can be used to find how many children
   ride the school bus? Choose all the correct answers.

   A. 28 + 32 = ?   B. 32 + 28 = ?

   C. 28 + 23 = ?   D. 23 + 28 = ?

Reflect On Your Learning

Lesson 5-5

Exit Ticket

Name _____

1. How do you decompose the addends in 25 + 44?

   A. 2 + 5 and 4 + 4   **B.** 20 + 5 and 40 + 4

   C. 20 + 4 and 40 + 5   **D.** 50 + 2 and 40 + 4

2. Decompose both addends by place value.

   $$56 \qquad + \qquad 23 = ?$$

   _____ + _____   _____ + _____

3. There are 37 fish in a pond. 18 more fish come. How many fish are in the pond now? Decompose both addends to find the sum. Show your work.

Reflect On Your Learning

Exit Ticket

Name _____

I. Which equation does the number line show?

| 0 | 5 | 10 | 15 | 20 | 25 | 30 | 35 | 40 | 45 | 50 | 55 | 60 | 65 | 70 | 75 | 80 | 85 | 90 | 95 | 100 |

| 34 | 47 |

A. $33 + 47 = 80$ **B.** $34 + 48 = 82$

C. $34 + 47 = 81$ **D.** $34 + 47 = 82$

2. What is the sum of $28 + 46$? Use the number line to help you.

| 0 | 5 | 10 | 15 | 20 | 25 | 30 | 35 | 40 | 45 | 50 | 55 | 60 | 65 | 70 | 75 | 80 | 85 | 90 | 95 | 100 |

$28 + 46 = $ _____

3. There are 28 reptiles and 33 birds at a zoo. How many reptiles and birds are at the zoo? Use the number line to help you.

| 0 | 5 | 10 | 15 | 20 | 25 | 30 | 35 | 40 | 45 | 50 | 55 | 60 | 65 | 70 | 75 | 80 | 85 | 90 | 95 | 100 |

Reflect On Your Learning

Exit Ticket

Name _____

Ia. How can you decompose 29 to help you find the sum of 29 + 44? Choose all the correct answers.

   A. $20 + 9$ **B.** $10 + 10 + 9$

   C. $20 + 4$ **D.** $6 + 20 + 3$

Ib. What is the sum?

   $29 + 44 =$ _____

2. How can you use the number line to add 65 + 27 by decomposing one addend? Fill in numbers to find the sum.

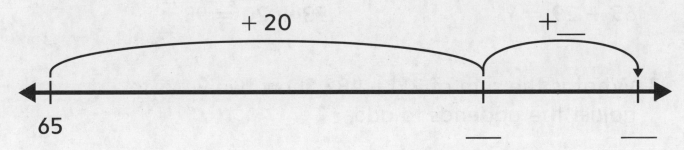

   $65 + 27 =$ _____

Reflect On Your Learning

Lesson 5-8

Exit Ticket

Name _____

1. Christy wants to adjust addends to find the sum of
   18 + 59. Which could Christy use to find the sum?
   Choose all the correct answers.

   A. 20 + 60   **B.** 20 + 57

   C. 15 + 60   **D.** 17 + 60

2. Match the equation with the correct friendly
   number equation.

   27 + 65 = ?   64 + 30 = 94

   74 + 19 = ?   30 + 62 = 92

   65 + 29 = ?   73 + 20 = 93

3. What is the sum of 27 + 48? Show two ways to
   adjust the addends to add.

Reflect On Your Learning

Exit Ticket

Name _____

1. Parker wants to add 18 + 39 + 22. How can you change the order of the addends to make friendly numbers?

   18 + _____ + _____

2. Write two ways the addends in the equation 24 + 17 + 29 + 22 = ? can be adjusted to find the sum. What is the sum?

3. Conrad has 19 sports books, 13 mystery books, 36 comic books, and 21 animal books. How many books does he have in all?

Reflect On Your Learning

Exit Ticket

Name _____

1. Trina has 48 stickers. She buys a package of 24 stickers. Then her friend gives her 15 stickers.

   a. Show the problem using base-ten shorthand.

   b. How many stickers does Trina have in all? Complete the equation.

   _____ + 24 + _____ = _____ stickers

2. Garrett picks 29 apples. Desmond picks 9 more apples than Garrett. How many apples do Garrett and Desmond pick in all?

Reflect On Your Learning

Performance Task

Name _____

School Store

The school store sells many items.

Item	Price
eraser	12 cents
pencil	28 cents
pen	45 cents
stickers	64 cents
highlighter	79 cents
crayon pack	87 cents
folder	32 cents
pencil sharpener	42 cents

Part A

Darius buys a pen and a folder. How much money does he spend? Draw base-ten blocks to show your work.

_____

Part B

Mrs. Hook buys 12 packs of stickers and 7 erasers. How many items does she buy? Use the number line to show your work.

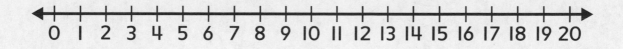

0 1 2 3 4 5 6 7 8 9 10 11 12 13 14 15 16 17 18 19 20

_____

Part C

The store has 38 crayon packs and 46 highlighters. How many crayon packs and highlighters does the store have? Decompose one addend to add.

_____

Part D

Georgia has 75 cents. Does she have enough money to buy an eraser, a pencil, and a folder? Show your work and explain your answer.

Part E

Iris wants to buy two things at the store. She plans to spend less than 75 cents. Which two items could she buy? Show your work and explain your answer.

Unit Assessment, Form A

Name _____

1. How can you decompose the second addend to make a 10 with the first addend? Choose the correct answer.

$$7 + 6 = ?$$

A. $1 + 5$ **B.** $2 + 4$ **C.** $3 + 3$ **D.** $0 + 6$

2. What is the sum?

$34 + 49 =$ _____

$49 + 34 =$ _____

3. Which equation does the number line show?

A. $28 + 64 = 93$ **B.** $29 + 65 = 94$

C. $29 + 64 = 83$ **D.** $29 + 64 = 93$

4. How can you adjust addends to find $48 + 34$? Choose all the correct answers.

A. $50 + 30$ **B.** $50 + 32$ **C.** $52 + 30$ **D.** $50 + 40$

5. How can you decompose by place value to add 57 + 24?

6. What is the sum of 8 + 6? Use the number line to help you.

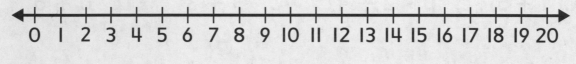

8 + 6 = _____

7. How can you decompose the second addend in 8 + 9 to make a double with the first addend?

A. 8 + 1   **B.** 7 + 2

C. 6 + 3   **D.** 5 + 4

8. Mel has 49 potatoes in her garden. She plants 23 more potatoes. How many potatoes does Mel have in her garden in all? Use the number line to help you.

```
├┼┼┼┼┼┼┼┼┼┼┼┼┼┼┼┼┼┼┼┼┼┼┼┼┼┼┼┼┼┤
0   5  10  15  20  25  30  35  40  45  50  55  60  65  70  75  80  85  90  95 100
```

Name _____

9. Emily buys 29 raspberries and 38 blueberries. How many berries does Emily buy in all? Use base-ten shorthand to solve the problem.

10. How can you decompose one addend to find the sum of $58 + 24$? Show your work.

11. Camden walks for 26 minutes and jogs for 35 minutes. Write two equations to find how many minutes Camden walks and jogs in all.

12. Kenna spends 32 minutes cleaning her room, 25 minutes dusting, and 16 minutes folding laundry. How many minutes does Kenna spend on chores in all?

13. Terry picks 58 red tomatoes and 27 yellow tomatoes. How many tomatoes does Terry pick in all? Decompose both addends to find the sum. Show your work.

14. On Monday, Grace spends 27 minutes swimming. On Tuesday, she spends 18 more minutes running than she does swimming on Monday. How many minutes does Grace spend swimming and running in all?

15. It snows 34 inches in January and 17 inches in February. Cassie says it snows 50 inches in January and February in all. How do you respond to Cassie? Use base-ten shorthand to explain your thinking.

16. Henry has 5 apple slices and 7 oranges slices. Use a doubles fact to help you find the total number of fruit slices Henry has. Show your work and explain your answer.

Unit Assessment, Form B

Name _____

1. How can you decompose the second addend to make a 10 with the first addend? Choose the correct answer.

$$9 + 7 = ?$$

A. $1 + 6$ **B.** $2 + 5$ **C.** $3 + 4$ **D.** $4 + 3$

2. What are the sums?

   $27 + 65 =$ _____

   $65 + 27 =$ _____

3. Which equation does the number line show?

| 34 | 45 |

A. $34 + 46 = 80$ **B.** $33 + 45 = 78$

C. $34 + 45 = 79$ **D.** $34 + 45 = 78$

4. How can you adjust addends to find $72 + 19$? Choose all the correct answers.

A. $71 + 20$ **B.** $70 + 20$ **C.** $70 + 22$ **D.** $70 + 21$

5. How can you decompose by place value to add 22 + 48?

22　　　　+　　　　48 =?

_____ + _____　　_____ + _____

6. What is the sum of 9 + 8? Use the number line to help you.

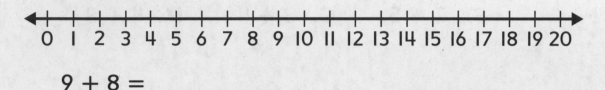

0 1 2 3 4 5 6 7 8 9 10 11 12 13 14 15 16 17 18 19 20

9 + 8 = _____

7. How can you decompose the second addend in 6 + 8 to make a double with the first addend?

A. 4 + 4　　　　　　　　**B.** 6 + 2

C. 5 + 3　　　　　　　　**D.** 7 + 1

8. Mr. Lee buys 26 thin paintbrushes and 35 thick paintbrushes. How many paintbrushes does Mr. Lee buy in all? Use the number line to help you.

0 5 10 15 20 25 30 35 40 45 50 55 60 65 70 75 80 85 90 95 100

9. A roller coaster has 32 red cars and 29 blue cars.
   How many cars does the roller coaster have in all?
   Use base-ten shorthand to solve the problem.

10. How can you decompose one addend to find the sum
    of 21 + 67? Show your work.

11. Florian folds 38 paper flowers. Max folds 25 paper
    flowers. Write two equations to find how many paper
    flowers Florian and Max fold in all.

12. Zofia counts 24 shells, 17 crabs, and 19 starfish at the
    beach. How many shells, crabs, and starfish does
    Zofia count in all?

13. A tree grows 44 centimeters the first year and 39 centimeters the second year. How many centimeters does the tree grow during the two years? Decompose both addends to find the sum. Show your work.

14. On Saturday, Jimin spends 26 minutes playing soccer. On Sunday, she spends 15 more minutes playing basketball than she does playing soccer on Saturday. How many minutes does Jimin spend playing soccer and basketball in all?

15. Aisha uses 29 star stickers and 23 flower stickers to make a craft. Aisha says she uses 50 stickers in all. How do you respond to Aisha? Use base-ten shorthand to explain your thinking.

16. There are 7 yellow birds and 8 brown birds at a park. Use a doubles fact to help you find the total number of birds at the park. Show your work and explain your answer.

How Ready Am I?

Name _____

1. What is the unknown number?

$10 - ? = 3$

A. 6   **B.** 7   **C.** 8   **D.** 13

2. Danielle has 5 hats. Fred has 17 hats. How many more hats does Fred have than Danielle?

A. 6 hats   **B.** 10 hats

C. 12 hats   **D.** 22 hats

3. Which number makes the equation true?

$70 - \square = 40$

A. 10   **B.** 20   **C.** 30   **D.** 40

4. Lila read 18 books over the summer. Connor read 7 fewer books than Lila. How many books did Connor read?

A. 9 books   **B.** 11 books

C. 12 books   **D.** 25 books

5. What is the unknown number?

$12 - ? = 3$

A. 9   **B.** 11   **C.** 14   **D.** 15

6. Logan is making 14 invitations for his party. He makes some of the invitations. He needs to make 8 more. How many invitations has Logan already made?

A. 6 invitations
B. 7 invitations
C. 14 invitations
D. 22 invitations

7. What is the difference of 30 − 10?

A. 2
B. 20
C. 30
D. 40

8. What is the unknown number?

? − 10 = 50

A. 10
B. 20
C. 40
D. 60

9. It took Tony 36 minutes to read a book. He read for 8 minutes on Monday, 13 minutes on Tuesday, and he read the rest of the book on Wednesday. How many minutes did Tony spend reading on Wednesday? Use the equation 8 + 13 + ? = 36.

A. 12 minutes
B. 14 minutes
C. 15 minutes
D. 57 minutes

10. What is the unknown number?

15 − 9 = ☐

A. 5
B. 6
C. 10
D. 24

Exit Ticket

Name

1. How can you show $18 - 9 = 9$ on the number line?

2. What subtraction equation is represented by the number line?

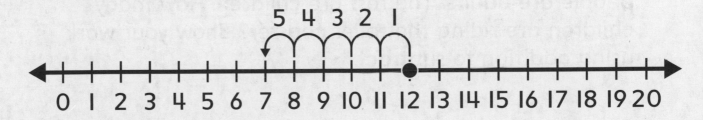

Equation:

3. Jeff is riding his bike on a trail that is 16 miles long. He has ridden 7 miles. How many miles are left?

Reflect On Your Learning

Exit Ticket

Name _____

1. How can you make a 10 to subtract? Fill in the numbers.

    15 − 6 = ?

    __ + __

    15 − 6 = ?

    15 − _____ = 10

    10 − _____ = 9

    15 − 6 = _____

2. There are 17 people riding a roller coaster. 8 of the people are adults. The rest are children. How many children are riding the roller coaster? Show your work using addition to subtract.

Reflect On Your Learning

Lesson 6-3

Exit Ticket

Name _____

1. Which subtraction equation do the base-ten blocks represent?

   A. $22 - 21 = 1$

   B. $34 - 12 = 22$

   C. $43 - 21 = 22$

   D. $43 - 12 = 31$

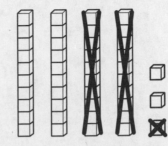

2. Garron's goal is to practice piano for 75 minutes each week. He has practiced for 34 minutes so far this week. How many more minutes does Garron need to practice this week to meet his goal?

   Draw base-ten shorthand to represent the problem. Then find the number of minutes Garron still needs to practice this week.

Reflect On Your Learning

Exit Ticket

Name _____

1. Match the equation to the base-ten blocks that represent how to find the difference.

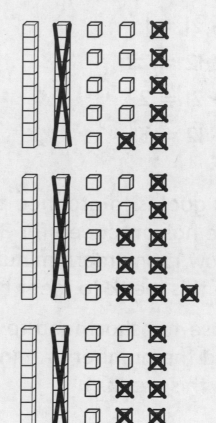

$36 - 19 = 17$

$35 - 18 = 17$

$35 - 16 = 19$

2. Elrich makes 42 pancakes for his friends and his family. His friends eat 29 pancakes. How many pancakes are left for Elrich's family?

Reflect On Your Learning

Exit Ticket

Name _____

1. Which equation matches the subtraction shown on the number line?

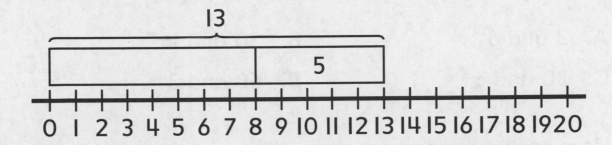

A. $13 + 5 = ?$ B. $13 - 5 = ?$

C. $18 - 5 = ?$ D. $18 - 13 = ?$

2. How can you use the number line to subtract? Fill in the difference.

$16 - 7 =$ _____

Reflect On Your Learning

Exit Ticket

Name _____

1. Asia is subtracting 74 − 36. She wants to decompose 36 to help her find the difference. Which pair of numbers can Asia use to subtract 74 − 36?

   A. 3 and 6   **B.** 30 and 16

   C. 30 and 6   **D.** 60 and 3

2. How can you decompose by place value to solve the problem?

   Jackson needs 82 dollars to buy a new bike. He has 24 dollars. How much money does Jackson need to buy the bike?

   $$82 - 24 = \underline{\quad}$$

   ___ + ___

Reflect On Your Learning

Exit Ticket

Name _____

1. Which way shows how to adjust the numbers to subtract? Choose all the correct answers.

   58 – 23

   A. 60 – 21

   B. 60 – 25

   C. 55 – 20

   D. 61 – 20

2. Adjust numbers to solve the problem. Show your work.

   Mark wins 81 tickets at an arcade. He trades in 48 tickets for a prize. How many tickets does Mark have left?

Reflect On Your Learning

Exit Ticket

Name _____

1. Match the subtraction equation to the addition equation that can be used to find its difference.

93 − 39 = ?   39 + ? = 65

65 − 39 = ?   ? + 65 = 93

93 − 65 = ?   39 + ? = 93

2. There are 52 people at a bowling alley. 27 people leave. How many people are left? Use an addition equation to solve the subtraction problem.

Reflect On Your Learning

Lesson 6-9

Exit Ticket

Name _____

1. Which equation represents the problem? Choose the correct answer.

   Imani has 76 newspapers to deliver. He has already delivered 49 newspapers. How many does he have left to deliver?

   A. $76 - 49 = ?$

   B. $76 + 49 = ?$

   C. $49 + 27 = ?$

   D. $49 - 72 = ?$

2. How can you represent and solve the problem? Fill in the equation and use any strategy to solve.

   It takes Andres 53 minutes to hike a trail. He has been hiking for 25 minutes. How many minutes does Andres have left to hike?

   _____ − _____ = _____

Reflect On Your Learning

Exit Ticket

Name

1. Write an equation that can be used to solve the problem.

   Carlos spends 73 minutes doing chores. He spends 36 minutes cleaning his room, 27 minutes raking leaves, and the rest of the time folding laundry. How many minutes does Carlos spend folding laundry?

   Equation:

2. How can you represent and solve the word problem? Use any strategy to solve.

   There are 64 people watching a juggler at a park. 25 adults and 18 children leave. How many people are still watching the juggler?

Reflect On Your Learning

Performance Task

Name

School Day

Ashanti makes a table showing how many minutes she spends in different activities throughout the school day.

Activity	Minutes
Math	72
Art	65
Writing	42
Lunch	34
Recess	15
Reading	45
Music	28

Part A

How many more minutes does Ashanti spend in Reading than Writing? Show your work.

Part B

What activity does Ashanti spend the most time doing? What activity does Ashanti spend the least time doing? How many fewer minutes does Ashanti spend in the activity she does least than the activity she does most? Show your work.

Part C

There are 38 minutes left in Art. How many minutes has Ashanti already spent in Art? Use the number line to show your work.

```
|++++|++++|++++|++++|++++|++++|++++|++++|++++|++++|++++|++++|++++|++++|++++|++++|++++|++++|++++|
0    5   10   15   20   25   30   35   40   45   50   55   60   65   70   75   80   85   90   95
```

Part D

There are 85 minutes left in the school day. Ashanti spends some of that time in Music, some of that time in Reading, and the rest of that time packing her things to leave. How much time does she have to pack her things? Show your work.

Part E

Does Ashanti spend more total time in Art and Music or in Writing and Reading? What is the difference in minutes? Show your work.

Unit Assessment, Form A

Name _____

1. What is the difference? Draw a base-ten shorthand representation. Then fill in the difference.

   61 − 28 = _____

2. Which way shows how to adjust the numbers to subtract? Choose all the correct answers.

   82 − 59

   A. 80 − 57   **B.** 80 − 61

   C. 81 − 60   **D.** 83 − 60

3. How can you use the number line to count back to subtract? Fill in the difference.

   15 − 7 = _____

4. What related addition equation can you use to find the difference?

   74 − 36 = ?

   Equation: _____ + _____ = _____

5. Mary knows she can make a 10 to help her subtract 14 − 8. What numbers can she decompose 8 into to make a 10 to subtract? Fill in the numbers.

She will decompose 8 into _____ and _____.

14 − 8 = _____

6. Miguel is subtracting 43 − 19. He wants to decompose 19 to help him find the difference. Which pair of numbers can Miguel use to subtract 43 − 19?

A. 1 and 9

B. 10 and 9

C. 10 and 19

D. 50 and 3

7. Match the subtraction equation to the addition equation that can be used to find its difference.

58 − 27 = ?   27 + ? = 85

72 − 58 = ?   ? + 27 = 72

85 − 27 = ?   58 + ? = 72

72 − 27 = ?   ? + 27 = 58

Name _____

8. Which equation matches the subtraction shown on the number line?

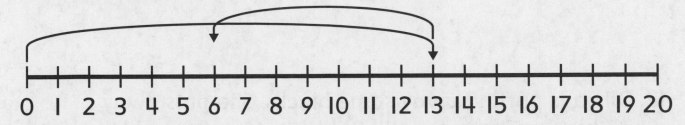

A. 13 − 7 = ?   **B.** 13 + 7 = ?

C. 13 + 8 = ?   **D.** 13 − 8 = ?

9. What is the difference? Draw a base-ten shorthand representation. Then fill in the difference.

57 − 35 = _____

10. How can you represent and solve the word problem? Use any strategy to solve.

There are 65 students eating lunch. There are 18 fewer students at recess than eating lunch. How many students are at recess?

11. How can you find the difference? Write step-by-step instructions explaining how to decompose a number to find the difference. Then fill in the difference.

$$73 - 34 = \underline{\hspace{3em}}$$

12. Olivia is putting away some blocks. She puts away 47 blocks. There are still 21 blocks to put away. How many blocks were out at the start? Complete the equation and explain how you can use a number line to subtract.

$$\underline{\hspace{3em}} - 21 = 47$$

13. How can you represent and solve the word problem? Use any strategy to solve.

Carina has 33 goats, 16 pigs, and some chickens on her farm. Carina has 91 animals on her farm. How many chickens does Carina have on her farm?

14. Armond eats 17 berries. He eats 9 blueberries. The rest are blackberries. How many blackberries does Armond eat? Show your work using addition to subtract.

Unit Assessment, Form B

Name _____

1. What is the difference? Draw a base-ten shorthand representation. Then fill in the difference.

   62 − 27 = _____

2. Which way shows how to adjust the numbers to subtract? Choose all the correct answers.

   81 − 38

   A. 80 − 39   B. 80 − 37

   C. 83 − 40   D. 79 − 40

3. How can you use the number line to count back to subtract? Fill in the difference.

   13 − 6 = _____

4. What related addition equation can you use to find the difference?

   75 − 59 = ?

   Equation: _____ + _____ = _____

5. Rosa knows she can make a 10 to help her subtract 15 − 7. What numbers can she decompose 7 into to make a 10 to subtract? Fill in the numbers.

She will decompose 7 into _____ and _____.

15 − 7 = _____

6. Ella is subtracting 34 − 16. She wants to decompose 16 to help her find the difference. Which pair of numbers can Ella use to subtract 34 − 16?

   A. 1 and 6

   B. 10 and 4

   C. 10 and 6

   D. 40 and 4

7. Match the subtraction equation to the addition equation that can be used to find its difference.

46 − 19 = ?   23 + ? = 64

64 − 23 = ?   ? + 19 = 64

46 − 23 = ?   19 + ? = 46

64 − 19 = ?   ? + 23 = 46

Name _____

8. Which equation matches the subtraction shown on
the number line?

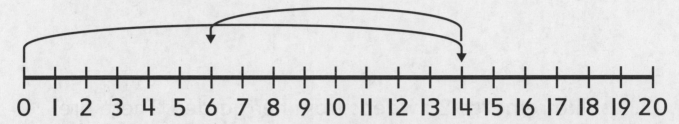

A. $14 - 7 = ?$ **B.** $14 + 7 = ?$

C. $14 + 8 = ?$ **D.** $14 - 8 = ?$

9. What is the difference? Draw a
base-ten shorthand representation.
Then fill in the difference.

$56 - 23 = $ _____

10. How can you represent and solve the word problem?
Use any strategy to solve.

There are 73 people sitting in booths at a restaurant.
There are 48 fewer people sitting at tables than in
booths. How many people are sitting at tables?

11. How can you find the difference? Write step-by-step instructions explaining how to decompose a number to find the difference. Then fill in the difference.

$85 - 39 =$ _____

12. Rishika removes 23 weeds from her garden. There are still 18 weeds to remove. How many weeds were in her garden at the start? Complete the equation and explain how you can use a number line to subtract.

_____ $- 18 = 23$

13. How can you represent and solve the word problem? Use any strategy to solve.

Carlos has 35 round buttons, 12 square buttons, and some triangle buttons. Carlos has 61 buttons in all. How many triangle buttons does Carlos have?

14. Levi can play 13 songs on the piano. He can play 5 jazz songs. The rest of the songs are classical. How many classical songs can Levi play? Show your work using addition to subtract.

Unit 7

How Ready Am I?

Name _____

1. $11 + 6 = ?$

   A. 14   **B.** 15   **C.** 16   **D.** 17

2. $15 + 4 = ?$

   A. 9   **B.** 16   **C.** 19   **D.** 20

3. $13 - 5 = ?$

   A. 7   **B.** 8   **C.** 9   **D.** 10

4. $18 - 7 = ?$

   A. 9   **B.** 10   **C.** 11   **D.** 12

5. Which object is shortest?

   A.

   B.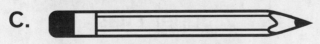

   C.

   D.

6. Which object is longest?

A. 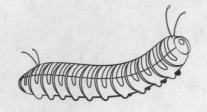　　　B.

C. 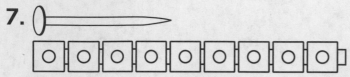　　　D.

About how many cubes long is the object?

7.

   A.　about 3 cubes　　　B.　about 4 cubes

   C.　about 5 cubes　　　D.　about 6 cubes

8. A.　about 5 cubes

    B.　about 6 cubes

    C.　about 7 cubes

    D.　about 8 cubes

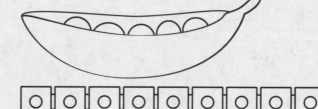

How many paper clips long is the object?

9.

   A.　4 paper clips　　　B.　5 paper clips

   C.　6 paper clips　　　D.　7 paper clips

10. A.　5 paper clips

    B.　6 paper clips

    C.　7 paper clips

    D.　8 paper clips

Exit Ticket

Name _____

1. What is the length of the flashlight?

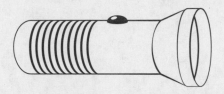

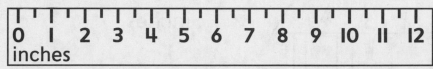

A. 4 inches **B.** 5 inches

C. 6 inches **D.** 7 inches

2. Matt saw a couple of insects in his yard. What is the length of each insect? Use an inch ruler to measure.

a.  b.

Reflect On Your Learning

Exit Ticket

Name _____

1. Match the unit you would use to measure the object to the object.

inches

feet

yards

2. What measuring tool would you use to measure the length of a hamster?

Reflect On Your Learning

Exit Ticket

Name _____

1. How can you compare the lengths? Fill in the equation.

   Abby's hair is 15 inches long. Val's hair is 8 inches long.

   _____ − _____ = _____

2. Which butterfly is longer? Fill in the equation.

   _____ − _____ = _____

   The top butterfly is _____ inches longer than the bottom butterfly.

3. Vera used 13 yards of yellow yarn and 6 yards of blue yarn to knit a scarf. How can you compare the lengths of the yellow and blue yarn? Fill in the equation.

   _____ − _____ = _____

Reflect On Your Learning

Exit Ticket

Name _____

1. Which two measures could be the length of a flute?

   A. 24 inches   **B.** 48 inches

   C. 2 feet   **D.** 4 feet

2. A tennis court is measured in feet and then in yards. Will the measurement of the length of a tennis court have more feet or yards?

   A. feet   **B.** yards

3. Jeremiah measures the length of his bicycle twice. The first time he uses feet and the second time he uses inches. Will Jeremiah's measurements have more feet or inches?

Reflect On Your Learning

Lesson 7-5

Exit Ticket

Name _____

1. Which everyday item can you use to estimate the length of a laptop computer?

    A. paper clip   B. golf club

2. How long is the instrument? Estimate the length.

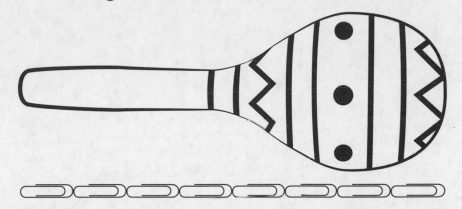

3. Calvin wants to estimate the length of his model airplane. What everyday item can he use to estimate the model airplane's length?

Reflect On Your Learning

Exit Ticket

Name

1. Suki is measuring the length of her driveway. Should she use a centimeter ruler or a meterstick?

2. Gabe is making a birdhouse. What are the lengths of the tools he is using? Use a centimeter ruler to measure.

   a.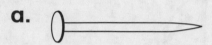

   b.

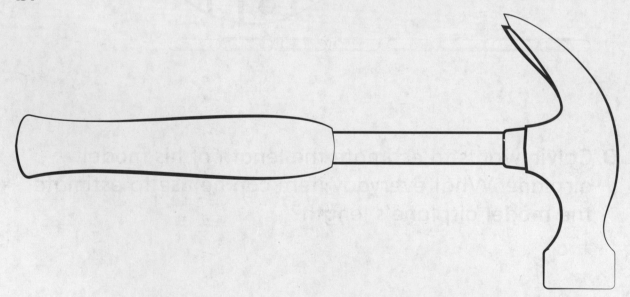

Reflect On Your Learning

Exit Ticket

Name _____

1. How can you compare the lengths? Fill in the equation.

   An ostrich's egg is 15 centimeters long. A chicken's egg is 6 centimeters long.

   _____ − _____ = _____

2. How much longer is the black feather than the white one?

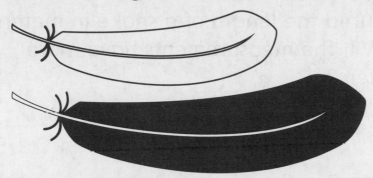

   _____

3. Mary's sidewalk is 15 meters long. John's sidewalk is 27 meters long. How much longer is John's sidewalk than Mary's sidewalk?

   _____

Reflect On Your Learning

Exit Ticket

Name _____

1. What measures could be the length of a rug? Choose all the correct answers.

   A. 2 centimeters   **B.** 2 meters

   C. 200 centimeters   **D.** 200 meters

2. A zookeeper measured the length of a snake in meters and centimeters. Will the measurements have more meters or centimeters?

   A. centimeters   **B.** meters

3. Graham measured the length of his bedroom in centimeters. Then he measured it in meters. Are there fewer centimeters or meters in his measurements?

Reflect On Your Learning

Exit Ticket

Name _____

1. Which unit would you use to measure the length of a watch?

   A. centimeter   **B.** meter

2. Which everyday items can you use to estimate the length of monkey bars? Choose all the correct answers.

   A. arm span   **B.** unit cube

   C. baseball bat   **D.** width of paper clip

3. How long is the whistle? Estimate the length.

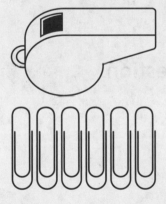

Reflect On Your Learning

Exit Ticket

Name _____

I. What operation would you use to solve the problem?

Henry is playing football. He runs the ball 16 yards. Then he runs the ball 9 yards. How many yards does Henry run the ball?

A. addition   **B.** subtraction

2. Roxanne's kite string is 55 feet long. Darin's kite string is 12 feet longer. How long is Darin's kite string?

a. How can you represent the problem with a drawing and an equation?

b. Solve the equation to answer the question.

Reflect On Your Learning

Lesson 7-11

Exit Ticket

Name _____

1. Preston's shoe is 13 centimeters long. His dad's shoe is 27 centimeters long. How much shorter is Preston's shoe than his dad's shoe?
Which equation can you use to solve the problem?

A. $27 + 13 = ?$ **B.** $27 - 13 = ?$

2. Mika has 24 meters of ribbon. He uses 11 meters of ribbon for party decorations. How many meters of ribbon does he have left?

a. How can you represent the problem with a drawing and an equation?

b. Use a number line to solve. Answer the question.

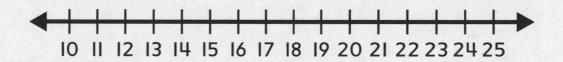

10 11 12 13 14 15 16 17 18 19 20 21 22 23 24 25

Reflect On Your Learning

Performance Task

Name _____

Carrie's Crafts

Carrie makes bracelets and scarves to sell at a craft fair.

Part A

How long is each bracelet?

child bracelet

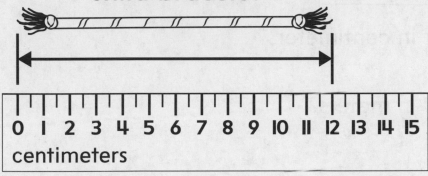

adult bracelet

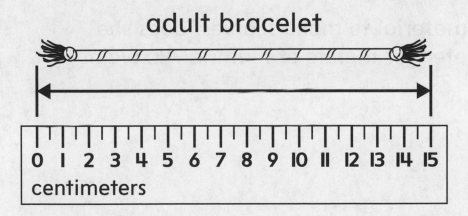

Part B

Which bracelet is longer?

adult   child

How much longer?

Part C

Carrie measures the scarves she makes.

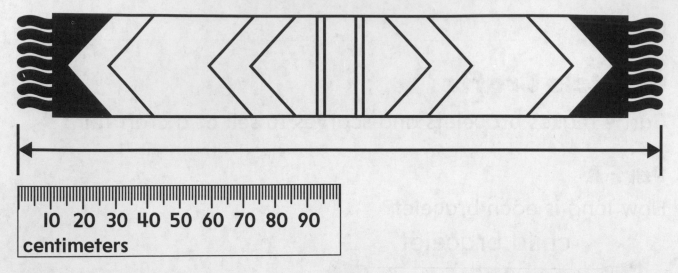

How long is the scarf in centimeters?

How long is the scarf in meters?

Part D

When Carrie buys material to make scarves, does she need fewer centimeters or meters of material? Explain.

Part E

Carrie wants to make 3 new scarves before the next craft fair. What length of material, in centimeters, will she need to buy to make the scarves? Write and solve a number sentence.

Unit Assessment, Form A

Name _____

1. How can you compare the lengths? Write the equation.

   Hannah's drone flies a distance of 32 yards.
   Josh's drone flies a distance of 24 yards.

   _____ − _____ = _____

2. Max measured the length of his canoe in centimeters. Then he measured it in meters. Are there more centimeters or meters?

3. What is the length of the object in inches? Use an inch ruler to measure.

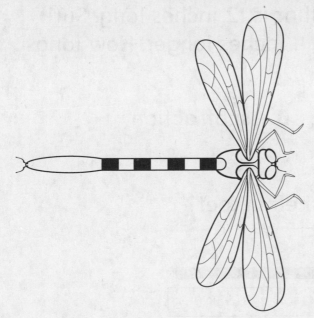

4. Which everyday item can you use to estimate the length of a green bean?

A. paper clip

B. math book

5. Which is the best tool to use to measure the length of a shampoo bottle?

A. ruler

B. yardstick

C. tape measure

6. What is the length of the grasshopper in centimeters? Use a centimeter ruler to measure.

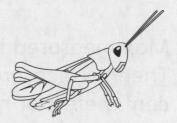

7. What operation would you use to solve the problem?

Miko has a coloring book that is 12 inches long. Suri has a coloring book that is 4 inches longer. How long is Suri's coloring book?

A. addition

B. subtraction

8. What is the length of the shelf in feet?

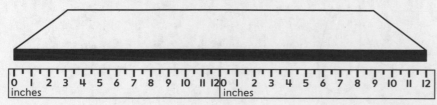

Name _____

9. Use a ruler to draw a line that is 6 inches long.

10. Which object is longer? Write the equation and the answer.

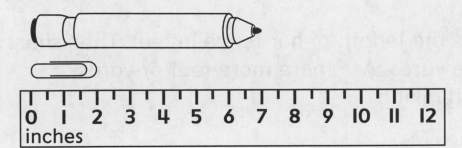

_____ − _____ = _____

The marker is _____ inches longer than the paper clip.

11. How can you compare the lengths? Write the equation.

Mary's cell phone is 14 centimeters long. Her laptop is 39 centimeters long.

_____ − _____ = _____

12. Which unit would you use to measure the length of a board game?

A. centimeter   **B.** meter

13. Will this toothpaste fit in a travel bag that is 15 centimeters long? Explain your thinking.

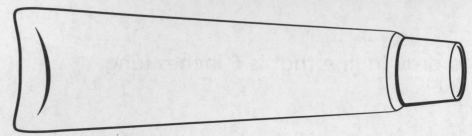

14. Lila measures the length of her fence in feet. Then she measures it in yards. Are there more feet or yards? Explain your thinking.

15. The cord on Anya's vacuum is 25 feet long. Her extension cord is 12 feet long. How long are the two cords combined? Explain.

16. Are there more centimeters or meters in two measurements of an object? Explain.

17. Cora has 45 yards of thread. She uses some of the thread to make string art. Cora has 27 yards of thread left. Explain how you can find how many yards of thread Cora used.

Unit Assessment, Form B

Name _____

I. How can you compare the lengths? Write the equation.

Ellarie's model rocket flies a distance of 46 yards. Tatum's model rocket flies a distance of 35 yards.

_____ − _____ = _____

2. Kinsey measured the length of her surfboard in centimeters. Then she measured it in meters. Are there more centimeters or meters?

3. What is the length of the object in inches? Use an inch ruler to measure.

4. Which everyday item can you use to estimate the length of a teeter-totter?

A. paper clip   **B.** math book

5. Which is the best tool to use to measure the length of a backyard?

    A. ruler   **B.** yardstick   **C.** tape measure

6. What is the length of the fish in centimeters? Use a centimeter ruler to measure.

7. What operation would you use to solve the problem?

Li buys a new hamster cage that is 8 inches longer than the hamster cage he already had. The length of the new hamster cage is 32 inches. What is the length of the hamster cage he already had?

    A. addition   **B.** subtraction

8. What is the length of the umbrella in feet?

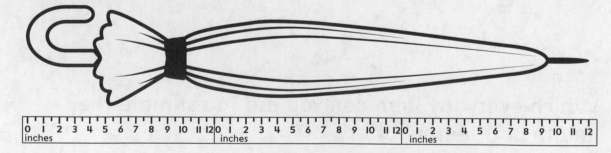

Name _____

9. Use a ruler to draw a line that is 7 inches long.

10. Which object is longer? Write the equation and
    the answer.

_____ − _____ = _____

The envelope is _____ inches longer than the pencil.

11. How can you compare the lengths? Write the equation.

Omar's football poster is 96 centimeters long. His
soccer poster is 72 centimeters long.

_____ − _____ = _____

12. Which unit would you use to measure the length of
    a shark?

A. centimeter   **B.** meter

13. Will the spoon fit in a section of a lunch box that is 14 centimeters long? Explain your thinking.

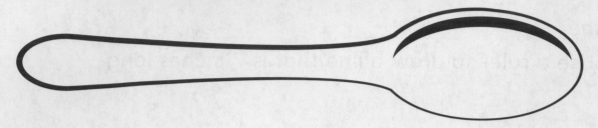

14. Jack measures the length of his fishing pole in feet. Then he measures it in yards. Are there more feet or yards? Explain.

15. Luna gallops 27 feet. Then she skips 51 feet. How long are the two distances combined? Explain.

16. Are there less centimeters or meters in two measurements of an object? Explain.

17. The length of Mark's garden is 27 feet. The length of Jin's garden is 11 feet shorter. Explain how you can find the length of Jin's garden.

Grade 2

Benchmark Assessment 2

Name _____

1. Look at the chalk.

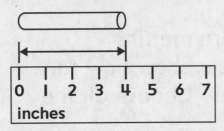

What is the length, in inches, of the chalk?

2. Look at the equation.

? − 23 = 52

Which is the unknown number?

A. 29   **B.** 31

C. 72   **D.** 75

3. Phillip made a red and yellow paper chain with 14 links. Some links are red and 9 links are yellow.

How many links on Phillip's paper chain are red?

4. Which number is represented by the expression 700 + 80 + 3?

A. 780

B. 783

C. 7,803

D. 700,803

5. Dylan has 7 muffins. Abel has 5 more muffins than Dylan. Which number sentences show how to find the number of muffins Abel has? Choose all the correct answers.

A. 7 − 5 = ?

B. ? − 5 = 7

C. 5 + ? = 7

D. 7 + 5 = ?

6. Andrew collects some marbles.

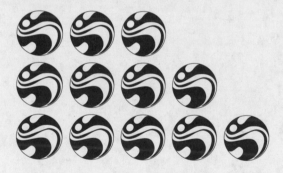

Which number sentence shows Andrew's marbles as the sum of two equal addends?

A. 3 + 3 = ?

B. 4 + 4 = ?

C. 5 + 5 = ?

D. 6 + 6 = ?

7. Elia picks 16 apples. She sells 8 of the apples. How many apples does Elia have left?

A. 6 apples

B. 7 apples

C. 8 apples

D. 9 apples

Name _____

8. Measure the spoon in inches and centimeters.

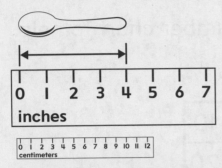

Which of these are correct estimates for the length of the spoon? Choose all the correct answers.

A. about 4 inches

B. about 10 inches

C. about 2 centimeters

D. about 4 centimeters

E. about 10 centimeters

9. Is there an even or odd number of triangles in the group? Circle Even or Odd for the group.

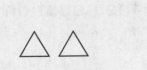

Even   Odd   Even   Odd   Even   Odd

10. Kaan counts by 1s starting from 537. Which digits are missing in the numbers he counts? Write the digits.

537, 53 _____, 539, 54 _____, 54 _____

11. What is the difference? Use the number chart to help.

52 − 29 = _____

1	2	3	4	5	6	7	8	9	10
11	12	13	14	15	16	17	18	19	20
21	22	23	24	25	26	27	28	29	30
31	32	33	34	35	36	37	38	39	40
41	42	43	44	45	46	47	48	49	50
51	52	53	54	55	56	57	58	59	60
61	62	63	64	65	66	67	68	69	70
71	72	73	74	75	76	77	78	79	80
81	82	83	84	85	86	87	88	89	90
91	92	93	94	95	96	97	98	99	100

12. Write the unknown number to complete the equation.

_____ = 53 + 28

13. Write the number in standard form.

5 hundreds, 4 tens, 2 ones = _____

700 + 10 + 9 = _____

three hundred six = _____

Name _____

14. How can you decompose one addend to add 38 + 53?
Fill in the missing numbers.

First, add the tens.

38 + _____ = _____

Then, add the ones.

_____ + 3 = _____

15. Renitha orders 8 yards of patterned fabric and
2 yards of solid fabric.

Which number line represents the total length of
fabric, in yards, that Renitha orders?

A.
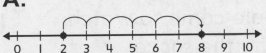

B.
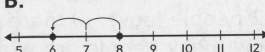

C.
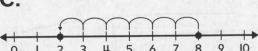

D.
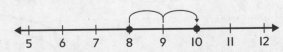

16. Nadia uses ten-frames to add 7 + 6.

Draw the missing counters to show how Nadia makes
a 10 to add.

17. There are 4 rows of desks in Mr. Burns' classroom. Each row has 5 desks.

Which shows the correct way to skip count to find the total number of desks in Mr. Burns' classroom?

A. 0, 10, 20, 30 = 30 desks

B. 10, 20, 30, 40 = 40 desks

C. 5, 10, 15, 20 = 20 desks

D. 10, 15, 20, 25 = 25 desks

18. There are 31 people watching a soccer game. 22 people leave. 10 more people come to watch. How many people are watching the soccer game now?

31

Fill in the missing numbers to find the number of people watching the soccer game now. Use the number line to help you.

31 − 22 = _____

_____ + 10 = _____

_____

How Ready Am I?

Name _____

1. What is the missing number?

   5, 10, _____, 20, 25

   A. 5   **B.** 10   **C.** 15   **D.** 20

2. Count by 5s. What are the next three numbers?

   45, 50, 55, 60, _____, _____, _____

   A. 61, 62, 63   **B.** 65, 70, 75

   C. 70, 75, 80   **D.** 70, 80, 90

3. 5 + 5 = ?

   A. 5   **B.** 10   **C.** 15   **D.** 20

4. 20 + 20 = ?

   A. 40   **B.** 50   **C.** 60   **D.** 80

5. Stef has 5 pens. She buys 5 more. Her mom gives her another 5 pens. How many pens does Stef have now?

   A. 5   **B.** 10   **C.** 15   **D.** 20

6. 10 + 10 + 10 = ?

   A. 20   **B.** 30   **C.** 40   **D.** 50

7. Jess picks 20 apples. Raul and Ashton each pick 10 apples. How many apples do they pick in all?

A. 40 **B.** 45 **C.** 50 **D.** 55

8. What time might you eat dinner?

A.

B.

C.

D.

9. Iva goes to bed at 8:30. Which shows Iva's bedtime?

A.

B.

C.

D.

10. Which time does the clock show?

A. half past 5:00 **B.** 5 o'clock

C. half past 6:00 **D.** 6 o'clock

Exit Ticket

Name _____

1. Match the coins with their value.

5¢

10¢

40¢

50¢

2. Fen has 90¢ in her wallet. All her coins are the same. What coins might Fen have? List all the possible coins she might have.

Reflect On Your Learning

Exit Ticket

Name _____

1. What is the value of the group of coins?

   A. 77¢   **B.** 78¢

   C. 82¢   **D.** 83¢

2. Bianca has 41¢. What combination of coins does she have if she has quarters, dimes, nickels, and pennies?

3. CJ buys a bag of pretzels for 2 quarters, 1 dime, 5 nickels, and 4 pennies. How much does the bag of pretzels cost?

Reflect On Your Learning

Exit Ticket

Name _____

1. What is the value of the group of bills?

$ _____

2. Charlotte is buying a video game for $38. Write two different combinations of bills she can pay with.

3. Jorge has one $20 bill, three $10 bills, two $5 bills, and one $1 bill. How much money does he have in all?

Reflect On Your Learning

Exit Ticket

Name _____

1. What time is shown on the analog clock?

_____ : _____

2. What time is shown on the analog clock? Circle the digital clock that shows the same time.

3. Cooper's hockey practice starts at 5:30 and ends at 6:15. What is another way to write each time?

Reflect On Your Learning

Exit Ticket

Name _____

1. Match the activity to the time it could take place.

   eating breakfast   9:00 p.m.

   putting on pajamas   3:30 p.m.

   playing outside   9:00 a.m.

2. What time of day would Regina most likely be doing homework? Write *a.m.* or *p.m.*

   4:30 _____

3. Cal is watching a movie. Would Cal most likely be watching a movie at 2:15 a.m. or 2:15 p.m.?

Reflect On Your Learning

Performance Task

Name _____

Landon's Lunch

Landon is buying lunch at school.

Part A

Landon needs some money for lunch today. His dad gives him these coins: 2 quarters, 2 dimes, 1 nickel, and 2 pennies. How much money does Landon's dad give him? Show how you counted.

Part B

Landon also has some dollar bills. How much money, in dollar bills, does he have? Show how you counted.

Part C

Landon spends 2 quarters on milk. How much money does he spend on milk?

Part D

Landon's lunchtime starts at the time shown on the clock. Write the time on the digital clock.

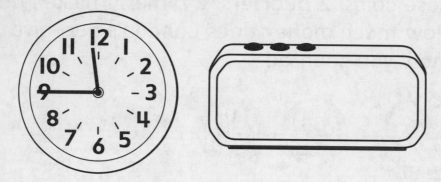

Part E

Do you think Landon's lunchtime starts in the a.m. or p.m.? Explain your answer.

Do you think Landon's lunchtime ends in the a.m. or p.m.? Explain your answer.

Unit Assessment, Form A

Name _____

1. Which of these coins is a dime?

A.   B.

C.   D.

2. Which groups of coins show the correct total?
   Choose all the correct answers.

A.  = 60¢

B. = 10¢

C.  = 30¢

D.  = 4¢

3. Franca has 3 nickels in her wallet and 4 nickels in her
   pocket. How much money does Franca have in all?

   A. 7¢   B. 19¢

   C. 35¢   D. 70¢

4. Andre puts the coins that were in his pocket on a table. What is the value of the group of coins?

_____ ¢

5. Mrs. Semple's children each have 50¢ in different coins. How many coins does each child have?

Polly has _____ quarters.

Macy has _____ nickels.

Isabel has _____ pennies.

Montel has _____ dimes.

6. Arthur has 64¢. What combination of coins could he have?

7. Jasper buys a toy car for 2 quarters, 2 dimes, 3 nickels, and 2 pennies. How much does the toy car cost?

8. What is the value of the group of bills?

$ _____

9. Jonah is buying some clothes for $56. Write two different combinations of dollar bills he can pay with.

10. Kalin has two $20 bills, three $10 bills, four $5 bills, and two $1 bills. How much money does she have in all?

11. Trent leaves for work at 8:05. Which clock shows the time Trent leaves for work?

A.

B.

C.

D.

12. Cecilia practices the flute from 3:30 to 4:15. What is another way to write each time?

13. When might the activity most likely happen? Match the activity to *a.m.* or *p.m.*

Go to the museum at 9:50

Meet at the movie theatre at 2:20   a.m.

Finish math class at 11:35

Go to soccer practice at 4:15

Eat a morning snack at 10:45   p.m.

Get home from school at 3:30

14. Cheri has some coins. She finds 3 dimes and 2 nickels in her couch. Now Cheri has 58¢. Cheri says she had 20¢ to begin with. How do you respond to Cheri?

15. Write one activity that has to take place in the a.m. and one activity that has to take place in the p.m. Explain your thinking.

Unit Assessment, Form B

Name _____

1. Which of these coins is a nickel?

A.

B.

C.

D.

2. Which groups of coins do *not* show the correct total? Choose all the correct answers.

A.  = 60¢

B. = 10¢

C.  = 30¢

D.  = 4¢

3. Gunnar finds 7 nickels in his car and 5 nickels in his couch cushions. How much money does Gunnar find in all?

A. 12¢

B. 50¢

C. 60¢

D. 70¢

4. Ron gets some coins from his grandma. What is the value of the group of coins?

_____ ¢

5. Marie puts 40¢ in three different envelopes using different coins. How many coins are in each envelope?

The red envelope has _____ nickels.

The blue envelope has _____ dimes.

The yellow envelope has _____ pennies.

6. Ali has 83¢. What combination of coins could she have?

7. Abram buys a hot dog for 2 quarters, 3 dimes, 2 nickels and 4 pennies. How much does the hot dog cost?

8. What is the value of the group of bills?

$_____

Name _____

9. Julia is buying some groceries for $47. Write two different combinations of bills she can pay with.

10. Marta has three $20 bills, two $10 bills, one $5 bill, and three $1 bills. How much money does she have in all?

11. Li has recess at 11:25. Which analog clock shows the time Li has recess?

A.

B.

C.

D.

12. Gage works on his homework from 6:30 to 7:15. What is another way to write each time?

13. When might the activity most likely happen? Match the activity to *a.m.* or *p.m.*

Go to a club meeting after school at 3:15

Get dressed for school at 8:10   a.m.

Make toast for breakfast at 7:50

Go to karate class at 4:30

Walk to the library at 11:20   p.m.

Read a bedtime story at 8:00

14. Ian has some coins. He finds 1 dime and 4 nickels on the sidewalk. Now Ian has 76¢. Ian says he had 36¢ to begin with. How do you respond to Ian?

15. Write two different activities that can take place in both the a.m. and p.m. Explain your thinking.

How Ready Am I?

Name _____

1. What number completes the pattern?

   451, 461, 471, 481, _____

   A. 482   **B.** 491

   C. 501   **D.** 581

2. What number completes the pattern?

   176, 276, 376, _____, 576

   A. 377   **B.** 386

   C. 476   **D.** 477

3. Which number do the base-ten blocks show?

   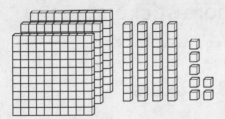

   A. 347   **B.** 437   **C.** 734   **D.** 743

4. Which of these shows a way to adjust numbers to add 25 + 18?

   A. 30 + 16   **B.** 23 + 20

   C. 20 + 13   **D.** 30 + 29

5. Which of these shows how to decompose both addends to add 54 + 63?

   A. $40 + 30 + 5 + 6$ **B.** $63 + 50 + 4 + 3$

   C. $60 + 30 + 50 + 40$ **D.** $50 + 60 + 4 + 3$

6. What is the sum of 30 + 6?

   A. 9   **B.** 24   **C.** 36   **D.** 44

7. What number is 10 more than 47?

   A. 37   **B.** 48   **C.** 57   **D.** 147

8. What is the sum of 58 + 34?

   A. 81   **B.** 88   **C.** 91   **D.** 92

9. Caleb has 19 stickers. He buys 35 more stickers. How many stickers does Caleb have now?

   A. 44 stickers   **B.** 53 stickers

   C. 54 stickers   **D.** 55 stickers

10. Ginny played tennis for 45 minutes on Saturday. She jogged for 22 minutes on Sunday. How many minutes did she exercise on Saturday and Sunday?

   A. 67 minutes   **B.** 68 minutes

   C. 77 minutes   **D.** 78 minutes

Exit Ticket

Name _____

1. What is the sum? Use the number line to show your thinking.

   545 + 10 = _____

2. What is the sum?

   693 + 10 = _____

   717 + 100 = _____

3. Emily reads the first 196 pages of a book. Then she reads 10 more pages. A week later, she reads 100 more pages and finishes the book. How many pages are in the book?

Reflect On Your Learning

Exit Ticket

Name _____

I. Which statement is true?

A. The total number of tens in the sum of 124 + 345 is 4.

B. The total number of tens in the sum of 237 + 561 is 7.

C. The total number of hundreds in the sum of 156 + 612 is 6.

D. The total number of hundreds in the sum of 418 + 570 is 9.

2. What is the sum? Use base-ten shorthand to solve.

167 + 131 = _____

3. The first grade classes at Green Elementary School collect 226 cans for a can drive. The second grade classes collect 353 cans. How many cans do the first and second grade classes collect?

Reflect On Your Learning

Exit Ticket

Name _____

1. Which equations need regrouping? Choose all the correct answers.

   A. 164 + 293 = ?   **B.** 245 + 324 = ?

   C. 317 + 271 = ?   **D.** 406 + 145 = ?

2. What is the sum? Use base-ten shorthand to show your work.

   159 + 282 = _____

hundreds	tens	ones

3. There are 178 red flowers and 186 yellow flowers in a garden. How many red and yellow flowers are in the garden?

Reflect On Your Learning

Exit Ticket

Name _____

1. What is the sum? Decompose both addends to solve.

   a.
   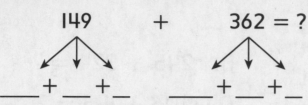
   $$149 \quad + \quad 362 = ?$$

   ___ + __ + _   ___ + __ + _

   b. Add hundreds: _____ + _____ = _____

   Add tens: _____ + _____ = _____

   Add ones: _____ + _____ = _____

   c. Add partial sums: _____ + _____ + _____ = _____

2. Garrett's class sold 285 raffle tickets last week and 476 raffle tickets this week. How many raffle tickets did Garrett's class sell during the two weeks? Decompose both addends to solve.

Reflect On Your Learning

Exit Ticket

Name _____

I. How can you decompose one addend to solve?
   Choose all the correct answers.

   534 + 128 = ?

   A. 534 + 100 + 2 + 8   **B.** 534 + 100 + 20 + 8

   C. 500 + 30 + 4 + 128   **D.** 500 + 3 + 4 + 128

2. What is the sum? Decompose one addend to solve.
   Use a number line to show your work.

   641 + 253 = _____

3. A hardware store sold 462 hammers and
   319 screwdrivers in a month. How many hammers
   and screwdrivers did the store sell in a month?
   Decompose one addend to solve.

Reflect On Your Learning

Exit Ticket

Name _____

I. How can you adjust the addends? Choose all the correct answers.

456 + 499

   A. 460 + 503   **B.** 460 + 495

   C. 457 + 500   **D.** 455 + 500

2. How can you adjust addends to find the sum? Fill in the numbers.

243 + 598 = ?

```
┌─────┐   ┌─────┐
│     │   │     │
└─────┘   └─────┘
   │         │
   ↓         ↓
```

_____ + _____ = _____

3. Sarah picks 202 blueberries and Maddy picks 327 blueberries. How many blueberries do Sarah and Maddy pick? Adjust the addends to solve.

Reflect On Your Learning

Exit Ticket

Name _____

1. What addition strategies are shown? Choose all the correct answers.

   176 + 242
   ↓
   200 + 40 + 2

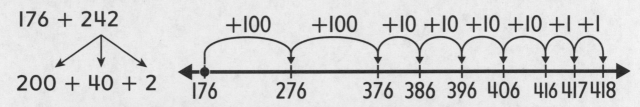

+100   +100   +10 +10 +10 +10 +1 +1

176   276   376 386 396 406 416 417 418

   A. adjust addends

   B. decompose both addends

   C. decompose one addend

   D. skip counting

2. Taylor found 417 + 339 = 756. Her work is shown. What addition strategy did Taylor use?

   417   +   339 = ?

   400 + 10 + 7   300 + 30 + 9

   400 + 300 = 700
   10 + 30 = 40
   7 + 9 = 16
   700 + 40 + 16 = 756

3. There are 502 pictures on Kaden's cell phone. He takes 184 pictures during a field trip. Now how many pictures are on Kaden's phone? Adjust the addends to solve.

Reflect On Your Learning

Performance Task

Name _____

Road Trip

Ezra and his family go on a road trip. They drive across the country. Ezra and his family live in Los Angeles.

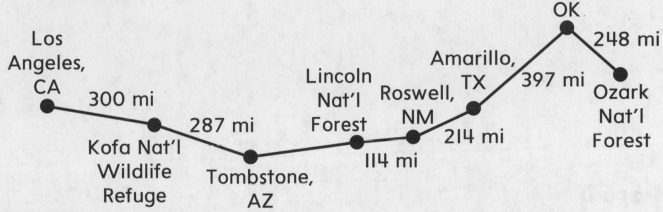

Part A

On Day 1, Ezra's family drives to the Wildlife Refuge. On Day 2, they drive to Tombstone. How many miles do they drive in the first two days? Show your thinking.

Part B

On Day 3, they drive to Lincoln Forest, which is 80 miles more than the distance they drive on Day 2. How far do they drive in all three days? Show your work.

Part C

On Day 4, Ezra and his family stop in Roswell, New Mexico. Then they drive to Amarillo, Texas. How many miles do they drive in all on Day 4? Use a different strategy to solve the problem. Explain why it is a useful strategy.

Part D

Ezra's family has two more days to travel. They plan to go to Pawhuska, OK the first day and to the Ozark National Forest the last day. How many miles will they drive on the last two days? Use a different strategy to solve the problem. Explain why it is a useful strategy.

Unit Assessment, Form A

Name _____

I. Is the statement true or false?
Circle the correct answer.

The total number of tens in the sum of 148 + 234 is 7.

True   False

2. How can you decompose one addend?
Choose all the correct answers.

$489 + 415$

A. $489 + 400 + 1 + 5$

B. $489 + 400 + 10 + 5$

C. $400 + 8 + 9 + 415$

D. $400 + 80 + 9 + 415$

3. What addition strategy is shown?

$$273 \quad + \quad 684 = ?$$

$200 + 70 + 3 \quad 600 + 80 + 4$

$200 + 600 = 800$
$70 + 80 = 150$
$3 + 4 = 7$
$800 + 150 + 7 = 957$

A. adjust addends

B. decompose both addends

C. decompose one addend

D. skip counting

4. Colleen read 361 pages in a book. She reads 10 more pages. How many pages has Colleen read in all?

  A. 362 pages   **B.** 371 pages

  C. 461 pages   **D.** 471 pages

5. How can you adjust the addends?
Choose all the correct answers.

  $297 + 602$

  A. $300 + 599$ **B.** $300 + 605$

  C. $295 + 600$ **D.** $299 + 600$

6. Which equations need regrouping?
Choose all the correct answers.

  A. $217 + 582 = ?$ **B.** $455 + 134 = ?$

  C. $346 + 471 = ?$ **D.** $263 + 648 = ?$

7. What addition strategy is shown?

  A. adjust addends

  B. decompose both addends

  C. decompose one addend

  D. skip counting

  $598 + 236 = 834$

  $\boxed{+2}\quad\boxed{-2}$

  $600 + 234 = 834$

8. Park rangers count 653 bison and 100 elk at a nature preserve. How many bison and elk do they count?

9. This month, Trent spends 254 minutes at soccer practice and 186 minutes at soccer games. How much time does Trent spend at soccer practice and soccer games this month? Decompose both addends to solve.

10. What is the sum? Decompose one addend to solve. Use a number line to show your work.

544 + 167 = _____

11. Stacy sells cars and trucks. She has 378 cars and 245 trucks for sale. What is the total number of cars and trucks Stacy has for sale?

12. What is the sum? Use base-ten shorthand to show your work.

432 + 325 = _____

hundreds	tens	ones

13. There are 194 children and 303 adults at a play. How many people are at the play? Explain two ways to adjust both addends.

14. What is the sum? Decompose one addend to solve. Use a number line to show your work.

$637 + 234 =$ _____

15. Tina adds $426 + 152$ by place value. She decomposes the addends as $400 + 2 + 6$ and $100 + 5 + 2$. Tina says the sum is 515. How do you respond to her?

16. Use two different addition strategies to find the sum of $705 + 283$. Which strategy do you think is more useful for this equation? Why?

Unit Assessment, Form B

Name _____

1. Is the statement true or false?
   Circle the correct answer.

   The total number of tens in the sum of 234 + 142 is 7.

   True   False

2. How can you decompose one addend?
   Choose all the correct answers.

   356 + 491

   A. 356 + 400 + 90 + 1

   B. 356 + 400 + 9 + 1

   C. 300 + 50 + 6 + 491

   D. 300 + 5 + 6 + 491

3. What addition strategy is shown?

   A. adjust addends   599 + 267 = 866

   B. decompose both addends   ┌─────┐ ┌─────┐
                                        │ +1 │ │ −1 │
   C. decompose one addend   └──┬──┘ └──┬──┘
                                           ↓   ↓
   D. skip counting   600 + 266 = 866

4. Xavier has 325 erasers. Then he buys 10 more erasers. How many erasers does Xavier have now?

   A. 326 erasers   **B.** 330 erasers

   C. 335 erasers   **D.** 425 erasers

5. How can you adjust the addends?
Choose all the correct answers.

198 + 404

   A. 200 + 406   **B.** 200 + 402

   C. 202 + 400   **D.** 194 + 200

6. Which equations need regrouping?
Choose all the correct answers.

   A. 271 + 558 = ?   **B.** 416 + 185 = ?

   C. 332 + 427 = ?   **D.** 245 + 603 = ?

7. What addition strategy is shown?

764   +   229 = ?   700 + 200 = 900

700 + 60 + 4   200 + 20 + 9   60 + 20 = 80
4 + 9 = 13
900 + 80 + 13 = 993

   A. adjust addends   **B.** decompose both addends

   C. decompose one addend   **D.** skip counting

8. A toy store has 348 dolls and 100 teddy bears. How many dolls and teddy bears does the toy store have?

Unit 9
Unit Assessment, Form B (continued)

Name _____

9. This month, Jana spends 257 minutes at softball practice and 166 minutes at softball games. How much time does Jana spend at softball practice and softball games this month? Decompose both addends to solve.

10. What is the sum? Decompose one addend to solve. Use a number line to show your work.

$584 + 135 =$ _____

11. Gavin has 273 stamps from the United States and 295 stamps from other countries. How many stamps does Gavin have in all?

12. What is the sum? Use base-ten shorthand to show your work.

$423 + 352 =$ _____

hundreds	tens	ones

13. There are 196 children and 502 adults at a fair. How many people are at the fair? Explain two ways to adjust both addends.

14. What is the sum? Decompose one addend to solve. Use a number line to show your work.

643 + 247 = _____

15. Andy adds 731 + 129 by place value. He decomposes the addends as 700 + 30 + 1 and 100 + 2 + 9. Andy says the sum is 842. How do you respond to him?

16. Use two different addition strategies to find the sum of 497 + 384. Which strategy do you think is more useful for this equation? Why?

How Ready Am I?

Name _____

1. Julia is counting back by 10s. What will be the next number Julia counts?

   44, 34, 24, 14, _____

   A. 0   **B.** 4   **C.** 10   **D.** 13

2. Abram is counting on by 10s. What will be the next number he counts?

   16, 26, 36, 46, _____

   A. 45   **B.** 47   **C.** 56   **D.** 65

3. Sari is counting back by 100s. What number is missing?

   937, 837, 737, _____, 537, 437

   A. 637   **B.** 647   **C.** 727   **D.** 736

4. What number do the base-ten blocks show?

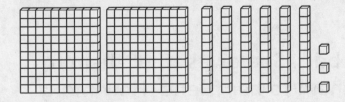

   A. 236   **B.** 263   **C.** 362   **D.** 632

5. Which number is 10 less than 83?

    A. 73   **B.** 82   **C.** 84   **D.** 93

6. Which number is 10 more than 67?

    A. 57   **B.** 68   **C.** 76   **D.** 77

7. Which of these shows how to decompose a number to find the difference of $67 - 38$?

    A. $67 - 20 - 10 - 1$ **B.** $67 - 27 - 7 - 1$
    C. $67 - 37 - 10 - 1$ **D.** $67 - 30 - 7 - 1$

8. Which equation shows the same difference as $58 - 37$?

    A. $58 - 40 = 18$ **B.** $60 - 40 = 20$
    C. $61 - 40 = 21$ **D.** $55 - 40 = 15$

9. Hannah has 54 stickers. Malcolm has 10 fewer stickers than Hannah. How many stickers does Malcolm have?

    A. 64 stickers   **B.** 55 stickers
    C. 53 stickers   **D.** 44 stickers

10. Luna makes 76 granola bars. She gives 10 of them to her neighbor and 10 of them to her friend. How many granola bars does she have left?

    A. 46 granola bars   **B.** 56 granola bars
    C. 66 granola bars   **D.** 67 granola bars

Lesson 10-1

Exit Ticket

Name _____

1. What is the difference?

    a. $362 - 10 =$ _____   b. $362 - 100 =$ _____

    c. $653 - 10 =$ _____   d. $653 - 100 =$ _____

2. Match the difference to the correct equation. Not all numbers will be used.

    $593 - 10 = ?$ 754

    $764 - 100 = ?$ 664

    $593 - 100 = ?$ 583

    $764 - 10 = ?$ 592

                                            493

3. There are 917 seats in the concert hall. There are 10 empty seats. How many seats are *not* empty in the concert hall?

Reflect On Your Learning

Exit Ticket

Name _____

1. Which equation is represented by the base-ten blocks?

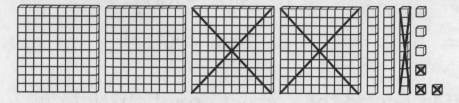

A. $436 - 113 = 323$

B. $436 - 231 = 205$

C. $436 - 213 = 223$

D. $436 + 213 = 223$

2. Mr. Smith's class collected 385 pennies and Miss Patton's class collected 164 pennies for a penny drive. How many more pennies did Mr. Smith's class collect than Miss Patton's class? Represent the problem using base-ten shorthand.

Reflect On Your Learning

Exit Ticket

Name _____

1. How can you decompose 346?

   A. 300 + 4 + 6   **B.** 300 + 40 + 6

2. How can you decompose by place value to find the difference? Show the subtraction on the number line.

   713 − 425 = _____

   425 = _____ + _____ + _____

3. Shawn has 304 shells and Callie has 132 shells. How many more shells does Shawn have than Callie? Decompose to find the difference. Show the subtraction on the number line.

Reflect On Your Learning

Exit Ticket

Name _____

1. Look at the equation $628 - 275 = ?$. Which equation is related? Choose the correct equation.

   A. $628 + 275 = ?$ **B.** $275 + ? = 628$

   C. $? - 275 = 628$ **D.** $275 + 628 = ?$

2. What is the difference? Use the number line to count on.

547   893

$893 - 547 =$ _____

3. Ethan's class has 354 recipe books. They sell 136 recipe books. How many recipe books do they have left? Write an equation. Use the number line to count on.

136   354

Reflect On Your Learning

Exit Ticket

Name _____

1. Which expressions need regrouping to subtract? Choose all the correct answers.

   A. 283 − 169   **B.** 416 − 204

   C. 657 − 321   **D.** 795 − 578

2. What is the difference? Use base-ten shorthand to show your work.

   342 − 117 = _____

3. Dylan puts 462 toy animals on the shelves of a zoo gift shop. Customers buy 134 toy animals. How many toy animals are still on the shelves?

Reflect On Your Learning

Exit Ticket

Name _____

1. Which need to be regrouped to subtract?

   435 − 246 = ?

   A. tens   **B.** hundreds   **C.** both

2. What is the difference? Use base-ten shorthand to show your work.

   321 − 168 = _____

3. A store has 512 cans of cat food and 379 cans of dog food. How many more cans of cat food than dog food does the store have?

Reflect On Your Learning

Exit Ticket

Name _____

1. How can you adjust the numbers to subtract? Choose all the correct answers.

   $449 - 153 = ?$

   A. $450 - 152$ **B.** $452 - 150$

   C. $450 - 154$ **D.** $446 - 150$

2. How can you adjust numbers to find the difference? Fill in the numbers.

   $728 - 396 =$ _____

   ```
   ┌─────┐   ┌─────┐
   │     │   │     │
   └──┬──┘   └──┬──┘
      ↓         ↓
   ____ - ____ = ____
   ```

3. Sefa practices piano for 805 minutes in April and 645 minutes in May. How many more minutes does Sefa practice piano in April than in May? What friendly equation can you use to solve?

Reflect On Your Learning

Lesson 10-8
Exit Ticket

Name _____

1. How can you adjust the numbers to find the difference?

   Choose all the correct answers. $352 - 146 = ?$

   A. $350 - 144$ **B.** $350 - 148$

   C. $348 - 150$ **D.** $356 - 150$

2. Which equations are related to $625 - 298 = ?$
   Choose all the correct answers.

   A. $625 + 298 = ?$ **B.** $625 - ? = 298$

   C. $298 + ? = 625$ **D.** $? - 298 = 625$

3. Write two ways you can decompose 537 to find the difference of $874 - 537$.

4. A clown has 451 balloons. The clown uses 148 balloons to make balloon animals. How many balloons does the clown have left? Explain what subtraction strategy you used.

Reflect On Your Learning

Exit Ticket

Name _____

1. Which equation can you use to represent the word problem?

   A tennis coach has 108 tennis balls. She buys 216 more tennis balls. How many tennis balls does the coach have in all?

   A. $216 - 108 = ?$ **B.** $216 - ? = 108$

   C. $108 + ? = 216$ **D.** $108 + 216 = ?$

2. Write an equation to represent the problem. Use any strategy to solve.

   Mr. Marley has 532 prizes in a box. Then 249 students each take one prize. Mr. Marley puts 175 more prizes in the box. How many prizes are in the box now?

Reflect On Your Learning

Unit 10

Performance Task

Name _____

Aquarium

Ida and her friends go to the aquarium for the day. They get there in the morning to wait in line.

Part A

There are 145 people in line. It is a special day so the first 100 people get in for free. How many people pay for their ticket? Explain how you solved the problem.

Part B

There are 256 people that work at the aquarium. There are 137 paid workers. The rest of the workers are volunteers. How many volunteers work at the aquarium? Represent and solve the problem.

Part C

The dolphins eat 765 fish that day. They eat 524 fish in the morning. How many fish do they eat the rest of the day? Show your work.

Part D

685 visitors eat lunch at the aquarium. 203 visitors eat dinner. How many more visitors eat lunch than dinner? Use a different strategy to solve the problem. Explain why it is a useful strategy.

Part E

There are 975 types of animals at the aquarium. 378 are mammals and 121 are reptiles. The rest are fish. How many types of animals at the aquarium are fish? Show your work.

Unit Assessment, Form A

Name _____

1. How can you decompose the bold number?

$314 - \mathbf{196} = ?$

A. $1 + 9 + 6$ **B.** $19 + 6$ **C.** $100 + 90 + 6$

2. Is regrouping needed to subtract? Circle Yes or No.

$573 - 258 = ?$

Yes   No

3. $602 - 10 = ?$

A. 702   **B.** 601

C. 592   **D.** 502

4. How can you adjust the numbers in $752 - 347$ to subtract? Choose all the correct answers.

A. $750 - 349$ **B.** $750 - 345$

C. $749 - 350$ **D.** $755 - 350$

5. What is the difference? Use base-ten shorthand to show your work.

$439 - 212 = ?$

6. How can you count on to subtract? Fill in the numbers on the number line and find the difference.

391 − 168 = _____

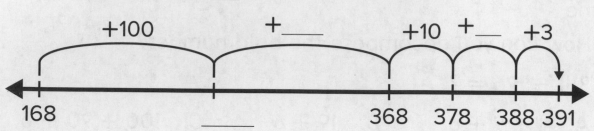

7. There are 523 shirts in a clothing store. During a sale, customers buy 100 shirts. How many shirts are left in the store?

8. Pablo needs to sell 400 raffle tickets. He has sold 268 tickets. How many does he still need to sell?

   a. How can you decompose by place value to solve the problem?

      268 = _____ + _____ + _____

   b. Show the subtraction on the number line.

9. Which needs decomposing to subtract 736 − 479?

   A. tens   **B.** hundreds   **C.** both

10. There are 287 people at a museum. Then 124 people leave the museum. How many people are at the museum now? Use base-ten shorthand to represent and solve the problem.

11. Write two ways to adjust the numbers in 853 − 549 to make friendlier numbers to subtract.

12. A sports store has 279 bikes. There are 10 bikes at the store with training wheels. How many bikes at the store do not have training wheels?

   A. 379 bikes   **B.** 289 bikes
   C. 269 bikes   **D.** 189 bikes

13. **a.** To count on to find the difference of 682 − 365, start at _____.

   b. To count back to find the difference of 682 − 365, start at _____.

14. There are 228 books in the returned book bin and 135 books on carts at a library that need to be put back on the shelves. By lunch time, 147 books have been put back on the shelves. How many books still need to be put back on the shelves?

15. Yuri is putting together a 650-piece puzzle. So far, Yuri has put together 322 pieces of the puzzle. Explain why regrouping is needed to find how many pieces Yuri still needs to put together.

16. A farmer grows 477 pumpkins. The farmer sells 268 of the pumpkins. How many pumpkins does the farmer have left? Choose a subtraction strategy to solve. Explain the subtraction strategy you used.

17. On Friday, 531 people go to the movie theater. On Saturday, 105 fewer people go to the movie theater than Friday. How many people go to the movie theater on Friday and Saturday in all? Explain your thinking.

Unit Assessment, Form B

Name _____

1. How can you decompose the bold number?

   476 − **238** = ?

   A. 200 + 30 + 8   **B.** 2 + 3 + 8   **C.** 23 + 8

2. Is regrouping needed to subtract? Circle Yes or No.

   562 − 351 = ?

   Yes   No

3. 703 − 10 = ?

   A. 603   **B.** 693

   C. 702   **D.** 713

4. How can you adjust the numbers in 354 − 148 to subtract? Choose all the correct answers.

   A. 350 − 144   **B.** 350 − 152

   C. 356 − 150   **D.** 352 − 150

5. What is the difference? Use base-ten shorthand to show your work.

   685 − 427 = ?

6. How can you count on to subtract? Fill in the numbers on the number line and find the difference.

892 − 558 = _____

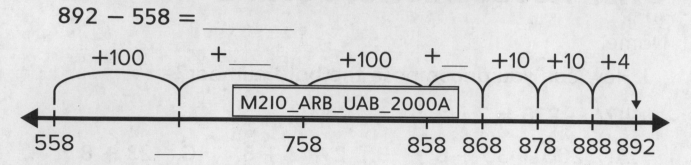

+100   +___   +100   +___   +10 +10 +4

558   ____   758   858 868 878 888 892

7. Vanessa has 446 old coins in her coin collection. She sells 100 of her coins. How many coins are left in Vanessa's coin collection?

8. Kyan needs to make 300 candles for a craft show. He has made 134 candles. How many candles does Kyan still need to make?

a. How can you decompose by place value to solve the problem?

134 = _____ + _____ + _____

b. Show the subtraction on the number line.

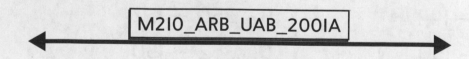

9. Which needs decomposing to subtract 565 − 271?

   A. tens   **B.** hundreds   **C.** both

10. Zoey wants to practice the violin for 360 minutes before her recital. She practices 118 minutes. How many more minutes does Zoey need to practice? Use base-ten shorthand to represent and solve the problem.

11. Write two ways to adjust the numbers in 652 − 347 to make friendlier numbers to subtract.

12. A cell phone store has 714 cell phones. There are 10 cell phones at the store with no camera. How many cell phones at the store have cameras?

   A. 604 cell phones   **B.** 614 cell phones

   C. 704 cell phones   **D.** 724 cell phones

13. **a.** To count on to find the difference of 493 − 268, start at _____.

   b. To count back to find the difference of 493 − 268, start at _____.

14. There are 354 pictures waiting to be framed at an art store. Customers bring in 289 more pictures to be framed. Over the weekend, 192 pictures are framed. How many more pictures need to be framed?

15. Suki is using 525 building bricks to make a castle. So far, Suki has put together 218 building bricks. Explain why regrouping is needed to find how many building bricks Suki still needs to put together.

16. A store has 436 watermelons. The store sells 129 of the watermelons. How many watermelons does the store have left? Choose a subtraction strategy to solve. Explain the subtraction strategy you used.

17. On Sunday, 602 people go to the zoo. On Monday, 264 fewer people go to the zoo than Sunday. How many people go to the zoo on Sunday and Monday in all? Explain your thinking.

Benchmark Assessment 3

Name _____

1. Look at the rectangle.

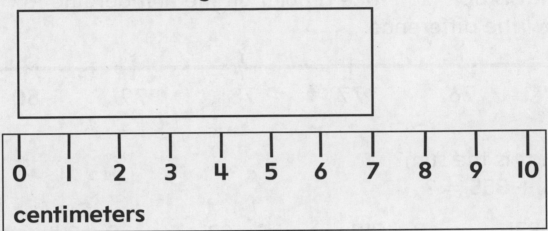

What is the length, in centimeters, of the rectangle?

2. Each equation has an unknown addend.

Choose Yes if the unknown addend is 100.

Choose No if the unknown addend is *not* 100.

	Yes	No
$536 + ? = 526$		
$? + 382 = 482$		
$608 + ? = 708$		
$? + 341 = 342$		

3. What unknown number makes the equation true?
Write the number.

$12 + 24 + 32 + 11 = $ _____

4. Timur plants 9 cherry trees and 8 apple trees.

How many trees does Timur plant?

5. What is 80 – 2? Place a point on the number line to show the difference.

75   76   77   78   79   80

6. What is the sum?
586 + 355 = ?

A. 831   **B.** 841   **C.** 931   **D.** 941

7. Eda looks at the clock in the morning.

What time does the clock show?

A. 10:02 a.m.   **B.** 10:10 a.m.

C. 2:10 p.m.   **D.** 2:50 p.m.

8. A store has 357 shirts to sell. They sell 120 shirts.

How many shirts does the store have left to sell?

Name _____

9. Which of these are equal to 400? Choose all the correct answers.

   A. 4 tens   **B.** 4 hundreds

   C. 40 tens   **D.** 40 ones

10. How can you decompose two addends to add 278 + 321? Fill in the missing numbers.

    First, decompose each addend into hundreds, tens and ones.

    278 = _____ + 70 + 8

    321 = 300 + _____ + 1

    Next, add the hundreds, add the tens, and add the ones.

    _____ + 300 = 500

    70 + _____ = _____

    8 + 1 = 9

    Finally, add the totals of the hundreds, tens, and ones together.

    278 + 321 = 500 + _____ + 9 = _____

11. Marla practices soccer 180 more minutes this week than Deon practices. Deon practices soccer for 235 minutes this week.

    How many minutes does Marla practice soccer this week?

12. Match the equation with its sum or difference. Not all numbers will be used.

	538
$438 - 10 = ?$	775
$675 + 100 = ?$	448
$438 + 100 = ?$	665
$675 - 10 = ?$	428
	575

13. Alanna has 5 dollar bills, 3 quarters, and 12 dimes in her piggy bank. She puts 6 more quarters in her piggy bank.

How much money does Alanna have in her piggy bank now?

A. $6.95   **B.** $7.01   **C.** $7.85   **D.** $8.45

14. Inez has 14 books. She organizes them in two rows.

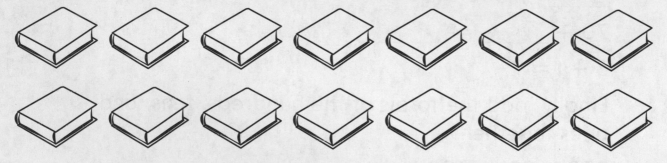

a. Is 14 an even or odd number?   Even   Odd

b. Make a double or near double to show how Inez organizes her books.

_____ books + _____ books = 14 books

Name _____

15. Look at the number chart.

1	2	3	4	5	6	7	8	9	10
11	12	13	14	15	16	17	18	19	20
21	22	23	24	25	26	27	28	29	30
31	32	33	34	35	36	37	38	39	40
41	42	43	44	45	46	47	48	49	50
51	52	53	54	55	56	57	58	59	60
61	62	63	64	65	66	67	68	69	70
71	72	73	74	75	76	77	78	79	80
81	82	83	84	85	86	87	88	89	90
91	92	93	94	95	96	97	98	99	100

Which statements are true about the column
that begins with the number 4? Choose all the
correct answers.

A. Every number in the column starts with a 4.

B. Every number in the column ends in a 4.

C. The numbers in the column show counting by 1s.

D. The numbers in the column show counting by 10s.

16. Norma has 7 pennies, 3 nickels, and 1 quarter. How much money does she have?

    A. 11 cents   **B.** 32 cents

    C. 47 cents   **D.** 62 cents

17. Which is *best* for measuring length in feet?

    A. 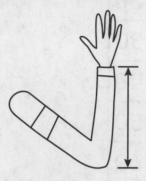   **B.**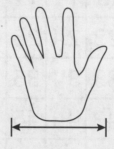

    C.    **D.**

18. Arnold has 8 more stickers than Hector. Arnold has 27 stickers. How many stickers does Hector have?

	--------27--------	
Arnold	_____	
Hector	_____	--8--	
	-----?-----	

Which equations match the word problem? Choose all the correct answers.

    A. $35 - 27 = 8$ **B.** $19 + 8 = 27$

    C. $27 - 8 = 19$ **D.** $27 + 8 = 25$

How Ready Am I?

Name _____

1. $5 + 3 = ?$

   A. 2   **B.** 7   **C.** 8   **D.** 9

2. $9 + 7 = ?$

   A. 2   **B.** 16   **C.** 17   **D.** 18

3. $7 - 2 = ?$

   A. 3   **B.** 4   **C.** 5   **D.** 9

4. $8 - 6 = ?$

   A. 2   **B.** 3   **C.** 4   **D.** 5

5. Which set of tally marks equals 5?

   A. || **B.** ||| **C.** |||| **D.** ||||

6. Which set of tally marks equals 8?

   A. |||| | **B.** |||| ||

   C. |||| ||| **D.** |||| ||||

7. Look at the tally chart. How many people chose almond as their favorite nut?

Favorite Nut	
Nut	**Tally**
Almond	꙰꙰꙰꙰꙰ ‖
Cashew	꙰꙰꙰꙰꙰ ꙰꙰꙰꙰꙰ ‖‖
Pecan	꙰꙰꙰꙰꙰ ‖‖

A. 6   **B.** 7

C. 8   **D.** 9

8. Look at the tally chart. How many people were asked the question: What is your favorite toy dinosaur party favor?

Toy Dinosaur Party Favors		
Type of Dinosaur		**Tally**
	Tyrannosaurus Rex	꙰꙰꙰꙰꙰
	Stegosaurus	꙰꙰꙰꙰꙰ ‖‖
	Velociraptor	꙰꙰꙰꙰꙰ ‖

A. 17   **B.** 18

C. 19   **D.** 20

9. How many members' answers are recorded in the picture graph?

Garden Club Flowers	
Tulip	🌷🌷🌷🌷🌷🌷🌷🌷🌷
Daisy	🌼🌼🌼🌼🌼
Marigold	🌼🌼🌼

Each picture = 1 member

A. 17   **B.** 18

C. 19   **D.** 20

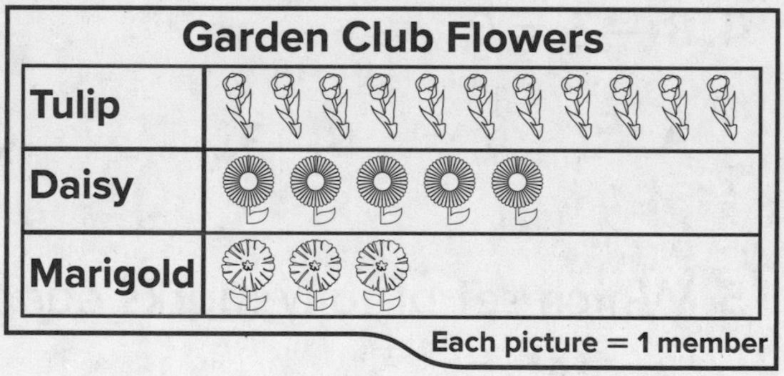

10. The picture graph shows the beads Mia used to make a necklace. How many green beads did she use?

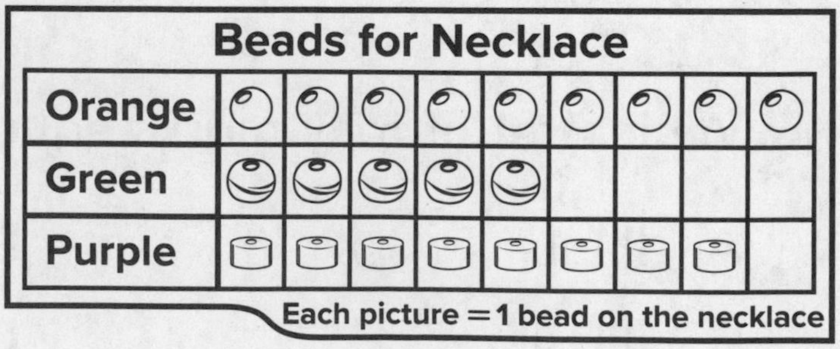

Beads for Necklace	
Orange	○○○○○○○○○
Green	⊙⊙⊙⊙⊙
Purple	⬭⬭⬭⬭⬭⬭⬭⬭

Each picture = 1 bead on the necklace

A. 4   **B.** 5

C. 8   **D.** 9

Lesson II-I

Exit Ticket

Name _____

I. Henry records the weather on different days. The results are in the picture graph. How many days are cloudy?

A. 4 days

B. 5 days

C. 6 days

D. 7 days

Weather						
Sunny	☀	☀	☀	☀	☀	
Cloudy	☁	☁	☁	☁	☁	☁
Rainy	🌧	🌧	🌧			
Windy	🌬	🌬				

Each picture = 1 day

2. The picture graph shows the animals in a tide pool. Which statements are true? Choose all the correct answers.

A. There are 4 starfish.

B. There are fewer crabs than starfish.

C. There are 15 animals in all.

D. There are more anemones than crabs.

Tide Pool Animals						
Starfish	★	★	★	★			
Crab	🦀	🦀	🦀	🦀	🦀	🦀	🦀
Anemone	🦔	🦔	🦔				
Jellyfish	🪼						

Each picture = 1 animal

Reflect On Your Learning

Exit Ticket

Name _____

1. The bar graph shows the favorite salad topping of customers at a restaurant. Which salad topping was chosen the most? The least?

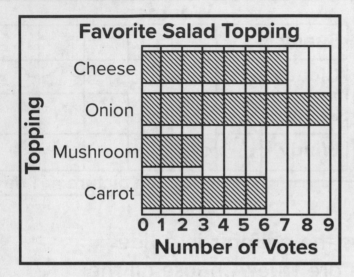

Favorite Salad Topping

2. How can you represent the data using a horizontal bar graph?

Favorite Picnic Foods	
Picnic Food	**Tally**
Fruit	ЖЖ |||
Salad	ЖЖ
Sandwiches	ЖЖ ||||
Vegetables	|||

Reflect On Your Learning

Exit Ticket

Name _____

Use the bar graph to answer the questions.

Favorite Berry

1. How many more people chose blueberry than raspberry?

   A. 3   **B.** 4   **C.** 5   **D.** 10

2. How many people chose strawberry and blackberry?

3. How many did not choose the most popular berry?

Reflect On Your Learning

Exit Ticket

Name _____

1. How can you make a tally chart to show the data?
   Maya measured the lengths of colored pencils in inches.

5 inches	6 inches
6 inches	5 inches
4 inches	6 inches
7 inches	7 inches

Length of Colored Pencil	
Length (inches)	Tally
4	
5	
6	
7	

2. Use the data to answer the questions.

   a. Sophie is making a tally chart of the data. How many rows should her tally chart have?

21 centimeters	24 centimeters
24 centimeters	21 centimeters
25 centimeters	25 centimeters
23 centimeters	22 centimeters

   b. How many tally marks go in the row for 22 centimeters?

Reflect On Your Learning

Exit Ticket

Name _____

Debra measured the lengths of her dolls. Use the data on the line plot to answer the questions.

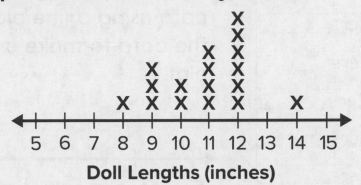

Doll Lengths (inches)

I. What is the most common length measured?

A. 9 inches

B. I0 inches

C. II inches

D. I2 inches

2. What are the least common lengths measured? Choose all the correct answers.

A. 8 inches

B. I0 inches

C. I4 inches

D. I5 inches

3. Debra got 3 more dolls. If she added their lengths to the line plot, how many measurements would be recorded?

Reflect On Your Learning

Exit Ticket

Name _____

I. Mila measured the lengths of bracelets she made.

○
18 centimeters
16 centimeters
14 centimeters
16 centimeters
○
15 centimeters
14 centimeters
18 centimeters
○ 16 centimeters

a. How can you represent the data using a line plot? Use the data to make a line plot.

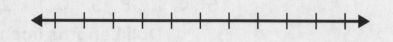

b. Write two questions using the line plot about bracelet lengths. Then answer the questions.

Reflect On Your Learning

Performance Task

Name _____

Ben's Graphs

Ben collects data with his friends.

Part A

Ben surveys his friends to find out their favorite colors. He makes a tally chart to record the data. How can you represent the data using a bar graph? Complete the bar graph to represent the data.

Favorite Color	
Color	Tally
Blue	IIII I
Yellow	IIII
Green	II
Red	IIII

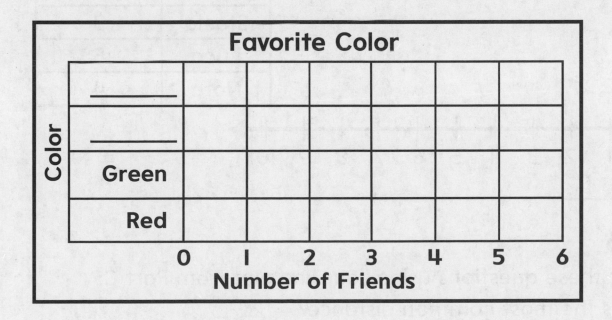

Part B

Answer these questions using your bar graph from Part A:

What color is the most popular?

How many more friends like blue than green?

How many friends like yellow or green best?

How many friends does Ben survey?
Show your work.

Part C

Ben and his friends test how far their toy cars roll after one push. They measure the distance the cars roll in feet. They record their data in the chart. How can you represent the data on a line plot? Complete the line plot to represent the data.

Distance Car Rolled	
Friends	Distance (feet)
Payton	7
Jamal	5
Addy	4
Oliver	4
Steela	3
Luca	5
Nora	4

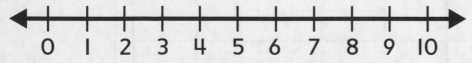

0 1 2 3 4 5 6 7 8 9 10

Part D

Answer these questions using your line plot from Part C:

What is the most common distance?

What is the least common distance?

How many measurements are shown?

Unit Assessment, Form A

Name _____

1. How can you make a tally chart to show the data?

Henry measured the lengths of the nails in his toolbox.

o
9 centimeters
5 centimeters
6 centimeters
8 centimeters
o
5 centimeters
9 centimeters
6 centimeters
o *5 centimeters*

Length of Nail	
Length (centimeters)	**Tally**
5	
6	
7	
8	
9	

2. Use the picture graph to answer the questions.

Ivan made this picture graph to record the animals he saw while camping.

Animals Seen While Camping

Deer — 🦌🦌🦌🦌🦌
Rabbit — 🐇🐇🐇
Raccoon — 🦝🦝
Squirrel — 🐿🐿🐿🐿🐿🐿🐿

Each picture = 1 animal

a. What animal did Ivan see the most?

b. How many deer did Ivan see?

3. Caleb measured the lengths of the items in his pencil box. Use the data on the line plot to answer the questions.

a. What is the most common length measured?

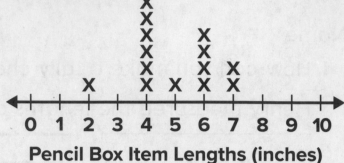

Pencil Box Item Lengths (inches)

b. How many measurements were recorded?

4. How can you represent the data using a line plot? Use the data to make a line plot.

Max measured the lengths of his toy trucks.

o	
	16 centimeters
	12 centimeters
	10 centimeters
	16 centimeters
o	18 centimeters
	16 centimeters
	10 centimeters
o	16 centimeters

Name

5. How can you represent the data using a picture graph?

Favorite African Animal	
Animal	**Tally**
Elephant	IIII
Giraffe	HHH I
Lion	HHH III
Zebra	III

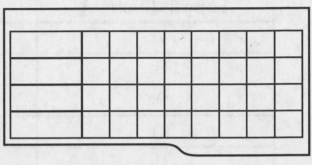

6. Use the bar graph to answer the questions.

a. What is the most popular vegetable?

b. How many more students chose carrots than broccoli?

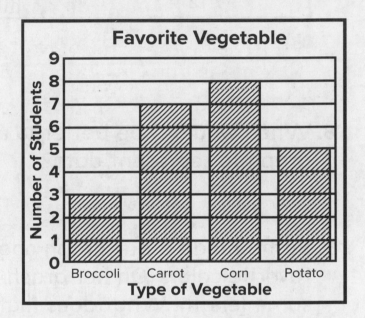

c. What is another observation you can make about this data?

7. Emilia measured the lengths of her hair ribbons and recorded the measurements in the tally chart. How might Emilia's tally chart change if she got 4 new ribbons that are each 14 inches long?

Length of Hair Ribbon	
Length (inches)	Tally
7	l
8	lll
9	
10	111
11	
12	llll

8. When would it be better to use a line plot than a bar graph to represent data?

9. Sasha made a bar graph about her classmates' favorite colors. On her graph, 3 of the bars were the same length. What does that mean?

Unit Assessment, Form B

Name _____

1. How can you make a tally chart to show the data?

A carpenter measured the lengths of the screws he had.

o	
	9 centimeters
	5 centimeters
	6 centimeters
	7 centimeters
o	5 centimeters
	6 centimeters
	6 centimeters
o	7 centimeters

Length of Screw	
Length (centimeters)	Tally
5	
6	
7	
8	
9	

2. Use the picture graph to answer the questions.

Kevin made the picture graph to record the animals he saw at a pond.

Animals Seen at Pond							
Duck	🦆	🦆	🦆	🦆	🦆	🦆	
Fish	🐟	🐟	🐟				
Frog	🐸	🐸	🐸	🐸			
Turtle	🐢						

Each picture = 1 animal

a. What animal did Kevin see the least?

b. How many frogs did Kevin see?

3. An art teacher measured the lengths of some markers. Use the data on the line plot to answer the questions.

a. What is the length of the shortest marker?

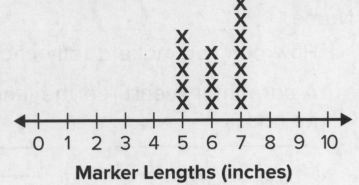

Marker Lengths (inches)

b. How many measurements were recorded?

4. How can you represent the data using a line plot? Use the data to make a line plot.

Max measured the lengths of his toy trains.

8 centimeters
10 centimeters
8 centimeters
6 centimeters
9 centimeters
8 centimeters
10 centimeters
8 centimeters

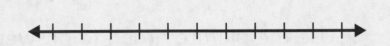

Name _____

5. How can you represent the data using a picture graph?

Favorite Bird	
Bird	**Tally**
Eagle	|||
Flamingo	⅂⅂⅂⅂⅂ ||
Parrot	⅂⅂⅂⅂⅂	
Penguin	⅂⅂⅂⅂⅂

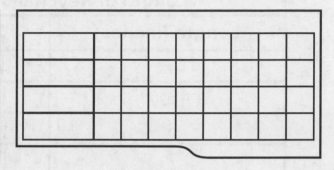

6. Use the bar graph to answer the questions.

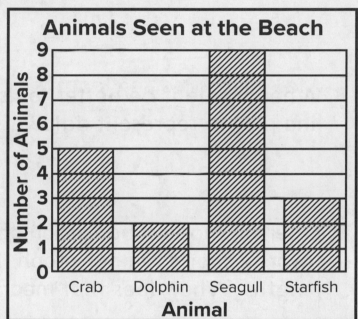

a. What animal was seen the most?

b. How many more crabs were seen than starfish?

c. What is another observation you can make about this data?

7. Gabriella measured the lengths of some keychains she made and recorded the measurements in the tally chart. How might Gabriella's tally chart change if she makes 5 more keychains that are each 3 inches long?

Length of Keychain	
Length (inches)	Tally
4	卌 ⅠⅠ
5	ⅠⅠⅠⅠ
6	Ⅰ
7	ⅠⅠ

8. When would it be better to use a bar graph than a line plot to represent data?

9. Emerson made a bar graph about her classmates' favorite sports. On her graph, the bars are different lengths. What does that mean?

How Ready Am I?

Name _____

1. Which is true for a square?

   A. It has 5 vertices.

   B. It has more sides than vertices.

   C. All its sides are the same length.

   D. 2 sides are short. 2 sides are long.

2. Which is true for a rectangle?

   A. It is a closed 2-dimensional shape.

   B. It has 5 vertices.

   C. All its sides are the same length.

   D. 3 sides are short. I side is long.

3. Which is a triangle?

   A.    B. 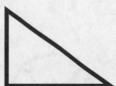   C.

4. Which is a hexagon?

   A.    B. 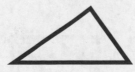   C.

5. Which has 4 vertices?

A.    B.    C.

6. Which is shaped like a cylinder?

A. 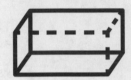   B. 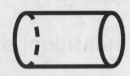   C.

7. Which is a rectangular prism?

A. 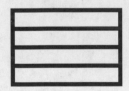   B.    C.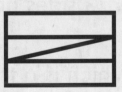

8. Which shows equal shares?

A.    B.    C.

9. Which shows halves?

A.    B.    C.

10. Which shows fourths?

A.    B.   C.

Exit Ticket

Name _____

1. Which shape has 6 sides, 6 angles, and 6 vertices?

   A. triangle   **B.** quadrilateral

   C. pentagon   **D.** hexagon

2. How many sides, angles, and vertices does the shape have?

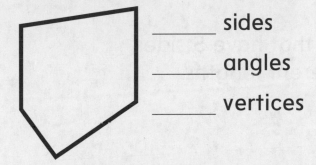

   _____ sides

   _____ angles

   _____ vertices

3. John received this letter in the mail. What shape is the envelope?

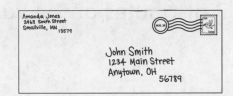

   A. hexagon   **B.** pentagon

   C. quadrilateral   **D.** triangle

Reflect On Your Learning

Exit Ticket

Name _____

1. Draw the shape. Then write the name.

   What shape has 6 sides, 6 angles, and all sides the same length?

2. Draw two different shapes that have 5 sides, 5 angles, and all sides different lengths.

3. Sara drew a vase. What are 3 attributes of the vase?

Reflect On Your Learning

Exit Ticket

Name _____

1. How many faces, edges, and vertices does the shape have? What is the shape?

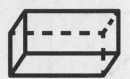
_____ faces   _____ vertices

_____ edges

This shape is a _____ .

2. Which shapes are spheres? Choose all the correct answers.

A.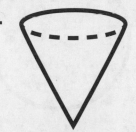

B.

C.

D.

3. Alex got a present for his birthday. What shape is the box?

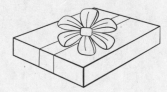

Reflect On Your Learning

Exit Ticket

Name _____

1. Which shapes are partitioned into equal shares? Choose all the correct answers.

 A.   B. 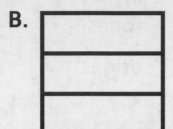   C.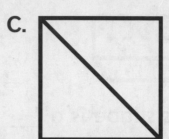

2. How can you partition the circle into 3 equal shares? Draw to show your work.

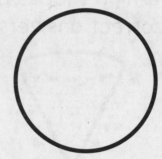

3. How can you partition the rectangle into 4 equal shares? Draw to show your work.

Reflect On Your Learning

Exit Ticket

Name _____

1. Which shows how to partition the same rectangle into fourths? Choose all the correct answers.

A.  B.  C.

2. How can you partition the squares into thirds? Show two different ways.

3. An apple slice is in the shape of a circle. Show how to partition the apple slice into halves in two different ways.

Reflect On Your Learning

Lesson 12-6

Exit Ticket

Name _____

1. How many rows, columns, and squares is the rectangle partitioned into?

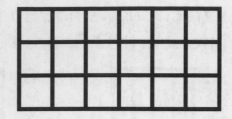

Rows: _____

Columns: _____

Write an equation to find the total number of squares.

Equation: _____

Total squares: _____

2. How can you partition the rectangle using equal-sized squares? Draw to show your work.

Total squares: _____

Reflect On Your Learning

Performance Task

Name _____

Art Class

Paul draws during art class.

Part A

Paul draws a shape with 4 sides and 4 angles. Draw two different shapes that Paul could draw. Circle the names of *all* the shapes Paul could draw.

hexagon   square

pentagon   trapezoid

rectangle   triangle

quadrilateral

Part B

Paul draws two circles. He divides one into halves and the other into fourths. Draw and label to show halves and fourths.

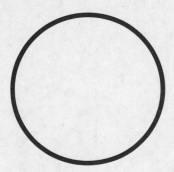   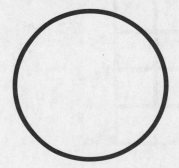

Part C

Paul draws two rectangles. He divides the rectangles into thirds. Draw two rectangles. Then show two different ways Paul can divide the rectangles into thirds. How are the equal shares related?

Part D

Paul's last drawing is a large square. He partitions the large square into rows and columns of equal-sized small squares. How many rows and columns are there? Write an equation that shows the total number of small squares.

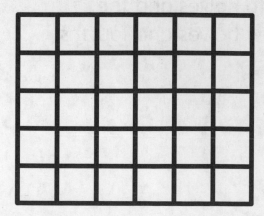

Unit Assessment, Form A

Name _____

1. What shape has 3 sides, 3 angles, and all sides different lengths? Draw the shape. Write its name.

2. Which shapes are partitioned into equal shares? Choose all the correct answers.

A.    B.    C. 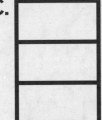   D.

3. Which shapes are pentagons? Choose all the correct answers.

A. 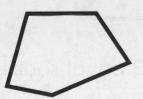   B.

C. 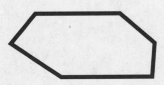   D.

4. How can you partition the rectangle using equal-sized squares? Draw to show your work.

Total squares: _____

5. How many faces, edges, and vertices does the shape have? What is the shape?

_____ faces

_____ edges

_____ vertices

This shape is a _____.

6. Which shows how to partition the same square into fourths? Choose all the correct answers.

A.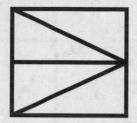

B.

C.

D.

7. How can you partition the circle into 3 equal shares? Draw to show your work.

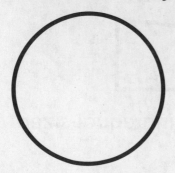

8. How many sides, angles, and vertices does the shape have?

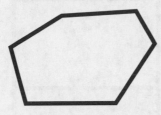

_____ sides

_____ angles

_____ vertices

9. How can you partition the rectangle into fourths? Show two different ways.

   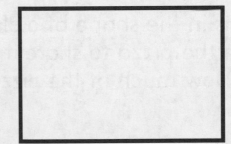

10. What shape has 1 base and 1 apex?

11. How many rows, columns, and squares is the rectangle partitioned into? Write an equation to find the total number of squares.

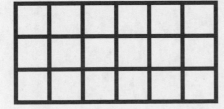

Rows: _____

Columns: _____

Equation: _____

_____

Total squares: _____

12. Cal drew the shape shown. What are three attributes of the shape?

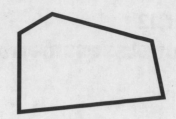

13. Kyle has a sandwich that is shaped like a square. He says he partitioned the sandwich into 2 equal shares so he can share the sandwich equally with his brother. How do you respond to Kyle?

14. A pizza is in the shape of a circle. Explain how to partition the pizza to share it equally between 4 people. How much of the pizza does each person get?

15. Brennan had the drink shown for lunch. What shape is it? Explain your thinking.

Unit Assessment, Form B

Name _____

1. What shape has 6 sides, 6 angles, and all sides different lengths? Draw the shape. Write its name.

2. Which shapes are partitioned into equal shares? Choose all the correct answers.

A. 　　B. 　　C. 　　D.

3. Which shapes are quadrilaterals? Choose all the correct answers.

A.

B.

C.

D.

4. How can you partition the rectangle using equal-sized squares? Draw to show your work.

Total squares: _____

5. How many faces, edges, and vertices does the shape have? What is the shape?

_____ faces

_____ edges

_____ vertices

This shape is a _____.

6. Which shows how to partition the same rectangle into thirds? Choose all the correct answers.

A.
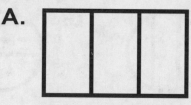

B.

C.
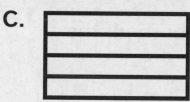

D.

7. How can you partition the square into 4 equal shares? Draw to show your work.

Name _____

8. How many sides, angles, and vertices does the shape have?

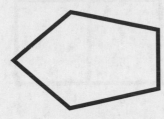

_____ sides

_____ angles

_____ vertices

9. How can you partition the circle into thirds? Show two different ways.

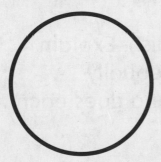 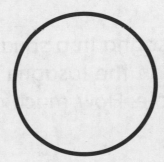

10. What shape has 2 bases and rolls?

11. How many rows, columns, and squares is the rectangle partitioned into? Write an equation to find the total number of squares.

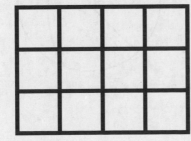

Rows: _____

Columns: _____

Equation: _____

_____

Total squares: _____

12. Mel drew this shape. What are three attributes of the shape?

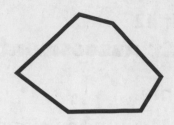

13. Carly has a cracker that is shaped like a rectangle. She says she partitioned the cracker into 2 equal shares so she can share the cracker equally with her sister. How do you respond to Carly?

14. Allie made lasagna in a square baking dish. Explain how to partition the lasagna to share it equally among 3 people. How much of the lasagna does each person get?

15. Mya is taking the ball shown to the beach. What shape is it? Explain your thinking.

Summative Assessment

Name _____

1. Which expression shows the correct way to find the sum of 180 + 52?

   A. 100 + 80 + 500 + 20   **B.** 100 + 80 + 50 + 2

   C. 100 + 80 + 50 + 20   **D.** 18 + 52 + 0

2. The bar graph shows the number of grapes children ate for a snack.

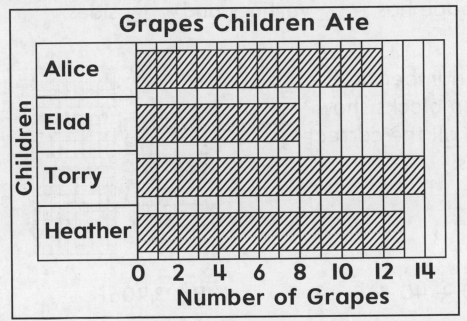

How many fewer grapes did Elad eat than Heather?

3. A rectangle is divided into rows and columns forming small squares of the same size.

How many squares are there in each row?

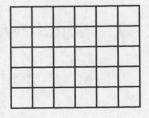

   A. 5   **B.** 6   **C.** 11   **D.** 30

4. Ali puts a 39 inch piece of wood next to a 25 inch piece of wood to make a garden border. What is the combined length of the two pieces of wood?

   A. 14 inches   **B.** 39 inches

   C. 54 inches   **D.** 64 inches

5. Complete the sentences to describe the shapes.

Each shape has _____ vertices and _____ sides.

6. Which number do the base-ten blocks show? Choose all the correct answers.

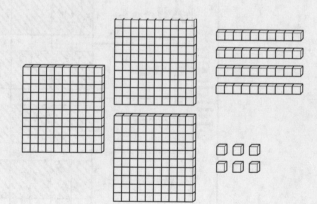

   A. 300 + 40 + 6   **B.** 3,406

   C. three thousand forty-six   **D.** 346

   E. three hundred forty-six

7. Which would be *best* measured using a meter stick?

   A. a nail   **B.** a door   **C.** a bean   **D.** a forest

Grade 2
Summative Assessment (continued)

Name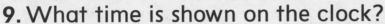

8. Luis starts at 338 and counts by Is. Fill in the missing numbers.

338, _____, _____, _____, _____, _____, 344

9. What time is shown on the clock?

A. 2:10

B. 1:50

C. 2:50

D. 10:10

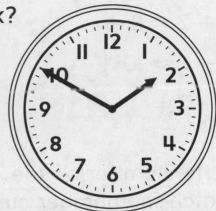

10. The monkeys at the zoo eat 5 bananas and 7 apples. How many bananas and apples do the monkeys eat?

A. 10   B. 11   C. 12   D. 13

11. Nikita adjusts numbers to subtract 57 – 38.

a. Which way shows how Nikita can adjust the numbers to subtract?

A. 60 – 40   B. 59 – 40

C. 55 – 40   D. 60 – 35

b. What is the difference?

57 – 38 = _____

12. Which number is two hundred two?

A. 22   B. 202   C. 220   D. 2,002

13. There are 174 apples at an apple farm. People pick 55 red apples and 36 green apples. How many apples are left at the apple farm?

A. 155   **B.** 138   **C.** 93   **D.** 83

14. Lucinda makes a 10 to add 8 + 7. Fill in the missing numbers to show how Lucinda adds.

8 + 7 = 8 + _____ + 5

8 + 7 = 10 + _____

8 + 7 = _____

15. Is the number on the number cube even or odd? Match the number cube to Even or Odd.

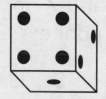   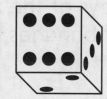

Even   Odd

16. What is the difference?

68 − 12 = _____

17. Which of these describes how the shape is divided into equal shares?

A. 2 halves   **B.** 3 thirds

C. 3 fourths   **D.** 4 fourths

18. Yin buys a muffin for 1 dollar and 50 cents and a bottle of juice for 2 dollars and 25 cents. How much money does Yin spend on the muffin and juice?

19. Which expression can be used to find the sum of 643 + 248?

   A. 640 + 250   **B.** 641 + 250

   C. 642 + 250   **D.** 645 + 250

20. The line plot shows the lengths of eggplants growing in Vince's garden. Which equation can you use to find out how many eggplants are shorter than 11 inches?

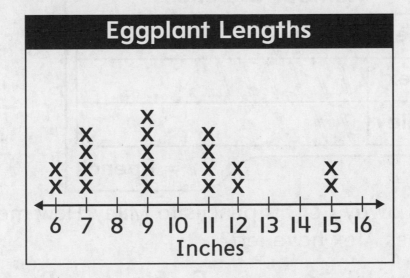

   A. 2 + 4 + 5 = 11   **B.** 6 + 7 + 9 = 22

   C. 11 + 4 = 15   **D.** 2 + 2 = 4

21. Which double can help to find the sum of 8 + 7?

A. $4 + 4 = 8$

B. $6 + 6 = 12$

C. $8 + 8 = 16$

D. $9 + 9 = 18$

22. Dana finds 8 more shells than Felix. Felix finds 7 shells. How many shells does Dana find?

23. What is the difference?

$821 - 10 =$ _____

24. The picture graph shows the number of pencils three students have.

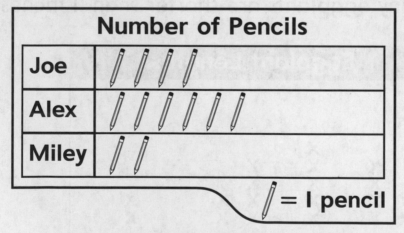

Alex gives away 1 of his pencils to Miley. How many pencils does Alex have left?

A. 2   B. 3   C. 5   D. 6